SpringerBriefs in Mathematics

T0172045

SpringerBriefs in Mathematics showcases expositions in all areas of mathematics and applied mathematics. Manuscripts presenting new results or a single new result in a classical field, new field, or an emerging topic, applications, or bridges between new results and already published works, are encouraged. The series is intended for mathematicians and applied mathematicians.

BCAM SpringerBriefs

BCAM *SpringerBriefs* aims to publish contributions in the following disciplines: Applied Mathematics, Finance, Statistics and Computer Science. BCAM has appointed an Editorial Board, who evaluate and review proposals.

Typical topics include: a timely report of state-of-the-art analytical techniques, bridge between new research results published in journal articles and a contextual literature review, a snapshot of a hot or emerging topic, a presentation of core concepts that students must understand in order to make independent contributions.

Please submit your proposal to the Editorial Board or to Francesca Bonadei, Executive Editor Mathematics, Statistics, and Engineering: francesca.bonadei@springer.com.

More information about this series at http://www.springer.com/series/10030

Peter Lindqvist

Notes on the Stationary
p-Laplace Equation

Peter Lindqvist
Department of Mathematical Sciences
Norwegian University of Science
and Technology
Trondheim, Norway

ISSN 2191-8198 ISSN 2191-8201 (electronic)
SpringerBriefs in Mathematics
ISBN 978-3-030-14500-2 ISBN 978-3-030-14501-9 (eBook)
https://doi.org/10.1007/978-3-030-14501-9

Library of Congress Control Number: 2019933174

This Springer imprint is published by the registered company Springer Nature Switzerland AG
The registered company address is: Gewerbestrasse 11, 6330 Cham, Switzerland

Preface

The most important partial differential equation of the second order is the celebrated Laplace equation. This is the prototype for linear elliptic equations. It is less well known that it also has a non-linear counterpart, the so-called p-Laplace equation, depending on a parameter p. The p-Laplace equation has been much studied during the last 50 years and its theory is by now well developed. Some challenging open problems remain, however. The p-Laplace equation is a degenerate or singular elliptic equation in divergence form. It deserves a treatise of its own, without any extra complications and generalizations. This is my humble attempt to write such a treatise. The interested reader may wish to consult the monograph *Nonlinear Potential Theory of Degenerate Elliptic Equations* by J. Heinonen, T. Kilpeläinen, and O. Martio when it comes to more advanced and general questions.

These notes were originally written up after my lectures at the Summer School in Jyväskylä in August 2005. Gradually, the text has been updated and extended, with new chapters about viscosity solutions and asymptotic mean values, among other topics. I am grateful to Juan Manfredi, who read the entire original text and contributed valuable comments and improvements. I thank Fredrik Hoeg, Marta Lewicka, Erik Lindgren, and Eero Ruosteenoja for help with proofreading.

Trondheim, Norway
January 2019

Peter Lindqvist

About This Book

This book in the BCAM SpringerBriefs series is a treatise on the p-Laplace equation. It is based on lectures by the author that were originally delivered at the Summer School in Jyväskylä, Finland, in August 2005 and have since been updated and extended to cover various new topics, including viscosity solutions and asymptotic mean values. The p-Laplace equation is a far-reaching generalization of the ordinary Laplace equation, but it is non-linear and degenerate $(p > 2)$ or singular $(p < 2)$. Thus it requires advanced methods. Many fascinating properties of the Laplace equation are, in some modified version, extended to the p-Laplace equation. Nowadays the theory is almost complete, although some challenging problems remain open.

Contents

About the Author

Peter Lindqvist is Professor of Mathematics in the Department of Mathematical Sciences, Norwegian University of Science and Technology, Trondheim, Norway. His research focus is analysis, including in particular partial differential equations and "non-linear potential theory".

Chapter 1
Introduction

The Laplace equation $\Delta u = 0$ or

$$\frac{\partial^2 u}{\partial x_1^2} + \frac{\partial^2 u}{\partial x_2^2} + \cdots + \frac{\partial^2 u}{\partial x_n^2} = 0$$

is the Euler-Lagrange equation for the Dirichlet integral

$$D(u) = \int_\Omega |\nabla u|^2 dx = \int \cdots \int_\Omega \left[\left(\frac{\partial u}{\partial x_1} \right)^2 + \cdots + \left(\frac{\partial u}{\partial x_n} \right)^2 \right] dx_1 \ldots dx_n.$$

If we change the square to a pth power, we have the integral

$$I(u) = \int_\Omega |\nabla u|^p dx = \int \cdots \int_\Omega \left[\left(\frac{\partial u}{\partial x_1} \right)^2 + \cdots + \left(\frac{\partial u}{\partial x_n} \right)^2 \right]^{\frac{p}{2}} dx_1 \ldots dx_n.$$

The corresponding Euler-Lagrange equation is

$$\operatorname{div}(|\nabla u|^{p-2} \nabla u) = 0.$$

This is the p-Laplace equation and the p-Laplace operator is defined as

$$\Delta_p u = \operatorname{div}(|\nabla u|^{p-2} \nabla u)$$

$$= |\nabla u|^{p-4} \left\{ |\nabla u|^2 \Delta u + (p-2) \sum_{i,j=1}^n \frac{\partial u}{\partial x_i} \frac{\partial u}{\partial x_j} \frac{\partial^2 u}{\partial x_i \partial x_j} \right\}.$$

Usually $p \geq 1$. At the critical points ($\nabla u = 0$) the equation is *degenerate* for $p > 2$ and *singular* for $p < 2$. The solutions are called p-harmonic functions.

© The Author(s), under exclusive license to Springer Nature Switzerland AG 2019
P. Lindqvist, *Notes on the Stationary p-Laplace Equation*,
SpringerBriefs in Mathematics, https://doi.org/10.1007/978-3-030-14501-9_1

There are several noteworthy values of p.

p = 1
$$\Delta_1 u = \text{div}\left(\frac{\nabla u}{|\nabla u|}\right) = -\text{H},$$

where H is the Mean Curvature Operator. In only two variables we have the familiar expression

$$\text{H} = \frac{u_y^2 u_{xx} - 2u_x u_y u_{xy} + u_x^2 u_{yy}}{(u_x^2 + u_y^2)^{\frac{3}{2}}}$$

The formula $\Delta_1 \varphi(u) = \Delta_1 u$ holds for general functions φ in one variable, indicating that solutions are determined by their level sets.

p = 2 We have the Laplace operator

$$\Delta_2 u = \Delta u = \sum_{i=1}^{n} \frac{\partial^2 u}{\partial x_i^2}.$$

p = n The borderline case. When n is the number of independent variables, the integral

$$\int_\Omega |\nabla u|^n dx = \int \cdots \int_\Omega \left\{\left(\frac{\partial u}{\partial x_1}\right)^2 + \cdots + \left(\frac{\partial u}{\partial x_n}\right)^2\right\}^{\frac{n}{2}} dx_1 \ldots dx_n$$

is conformally invariant. The n-harmonic equation $\Delta_n u = 0$ in n variables is therefore invariant under Möbius transformations. For example, the coordinate functions of the inversion (a Möbius transformation)

$$y = a + \frac{x-a}{|x-a|^2}$$

are n-harmonic. The borderline case is important in the theory of quasiconformal mappings.

p = ∞ As $p \to \infty$ one encounters the equation $\Delta_\infty u = 0$ or

$$\sum_{i,j=1}^{n} \frac{\partial u}{\partial x_i} \frac{\partial u}{\partial x_j} \frac{\partial^2 u}{\partial x_i \partial x_j} = 0.$$

This is the infinity Laplace equation. It has applications for *optimal* Lipschitz extensions and has been used in image processing.

The case $p > 2$ is *degenerate* and the case $1 < p < 2$ is *singular*.[1] In the classical theory of the Laplace equation several main parts of mathematics are joined in
a fruitful way: Calculus of Variations, Partial Differential Equations, Potential Theory, Function Theory (Analytic Functions), not to mention Mathematical Physics
and Calculus of Probability. This is the strength of the classical theory. It is very
remarkable that the p-Laplace equation occupies a similar position, when it comes
to non-linear phenomena. Much of what is valid for the ordinary Laplace equation
also holds for the p-harmonic equation, except that the Principle of Superposition
is naturally lost. Even the Mean Value Formula holds, though only infinitesimally
in small balls. A non-linear potential theory has been created with all its requisites:
p-superharmonic functions, Perron's method, barriers, Wiener's criterion and so on.
In the complex plane a special structure related to quasiconformal mappings appears.
Last but not least, the p-harmonic operator appears in physics: rheology, glacelogy,
radiation of heat, plastic moulding etc. Some advances indicate that even the Brownian motion has its counterpart and a mathematical game "Tug of War" leads to
the case $p = \infty$. For finite p a sophisticated stochastic game is a counterpart to the
Brownian motion.

Needless to say, the equation $\Delta_p u = 0$ has numerous generalizations. For example, one may start with variational integrals like

$$\int |\nabla u|^p \omega\, dx\,,\quad \int |\nabla u(x)|^{p(x)}\, dx,$$

$$\int \left| \sum_{i,j=1}^{n} a_{ij} \frac{\partial u}{\partial x_i} \frac{\partial u}{\partial x_j} \right|^{\frac{p}{2}} dx,$$

$$\int \left(\left|\frac{\partial u}{\partial x_1}\right|^p + \cdots + \left|\frac{\partial u}{\partial x_n}\right|^p \right) dx$$

and so on. The non-linear potential theory has been developed for rather general
equations

$$\operatorname{div} \mathbf{A}_p(x, \nabla u) = 0.$$

---

[1] The terminology comes from fluid dynamics. An equation of the type

$$\frac{\partial v}{\partial t} = \operatorname{div}(\rho(|\nabla v|)\nabla v)$$

is called singular, when it so happens that the diffusion $\rho = \infty$ and degenerate when $\rho = 0$. Notice
also the expansion

$$\rho(|\nabla v|) = |\nabla v|^{p-2} \left(c_1 + c_2|\nabla v| + c_2|\nabla v|^2 \cdots \right).$$

However, one may interpret Polya's Paradox[2] as indicating that *the special case is often more difficult than the general case*. In these lecture notes I resist the temptation of including any generalizations. Thus I stick to the pregnant formulation $\Delta_p u = 0$.

The p-harmonic operator appears in many contexts. A short list is the following.

- The non-linear eigenvalue problem

$$\Delta_p u + \lambda |u|^{p-2} u = 0$$

- The p-Poisson equation

$$\Delta_p u = f(x)$$

- Equations like

$$\Delta_p u + |u|^\alpha u = 0,$$

  which are interesting when the exponent α is "critical".
- Parabolic equations like

$$\frac{\partial v}{\partial t} = \Delta_p v,$$

  where $v = v(x, t) = v(x_1, \ldots, x_n, t)$
- So-called p-harmonic maps $\mathbf{u} = (u_1, u_2, \ldots, u_n)$ minimizing the "p-energy"

$$\int |D\mathbf{u}|^p dx = \int \left\{ \sum_{i,j} \left(\frac{\partial u_j}{\partial x_i} \right)^2 \right\}^{\frac{p}{2}} dx,$$

  perhaps with some constraints. A system of equations appears.
- The fractional p-Laplace operator

$$(-\Delta_p)^s u(x) = c_{p,s} \int_\Omega \frac{|u(y) - u(x)|^{p-2} (u(y) - u(x))}{|x - y|^{n+sp}} dy.$$

These additional topics are very interesting but cannot be treated here.

The reader is supposed to know some basic facts about L^p-spaces and Sobolev spaces, especially the first order spaces $W^{1,p}(\Omega)$ and $W_0^{1,p}(\Omega)$. The norm is

$$\|u\|_{W^{1,p}(\Omega)} = \left\{ \int_\Omega |u|^p dx + \int_\Omega |\nabla u|^p dx \right\}^{\frac{1}{p}}.$$

---

[2]"The more ambitious plan may have more chances of success", G. Polya, How to Solve It, Princeton University Press, 1945.

Ω is always a domain (= an open connected set) in the n-dimensional Euclidean space $\mathbf{R}^n$. Text books devoted entirely to Sobolev spaces are no good for our purpose. Instead we refer to [GT, Chap. 7], which is much to the point, [G, Chap. 3] or [EG]. The reader with an apt to estimates will enjoy the chapter "Auxiliary Propositions" in the classical book [LU].

Chapter 2
The Dirichlet Problem and Weak Solutions

The natural starting point is a Dirichlet integral

$$I(u) = \int_{\Omega} |\nabla u|^p dx \tag{2.1}$$

with the exponent p, $1 < p < \infty$, in place of the usual 2. Minimizing the integral among all admissible functions with the same given boundary values, we are led to the condition that the first variation must vanish, that is

$$\int_{\Omega} \langle |\nabla u|^{p-2} \nabla u, \nabla \eta \rangle dx = 0 \tag{2.2}$$

for all $\eta \in C_0^{\infty}(\Omega)$. This is the key to the concept of weak solutions. Under suitable assumptions this is equivalent to

$$\int_{\Omega} \eta \operatorname{div}(|\nabla u|^{p-2} \nabla u) dx = 0. \tag{2.3}$$

Since (2.3) has to hold for all test functions η, we must have

$$\Delta_p u \equiv \operatorname{div}(|\nabla u|^{p-2} \nabla u) = 0 \tag{2.4}$$

in Ω. In other words, the p-Laplace equation is the *Euler-Lagrange equation* for the variational integral $I(u)$.

It turns out that the class of classical solutions is too narrow for the treatment of the aforementioned Dirichlet problem. (By a classical solution we mean a solution having continuous second partial derivatives, so that the equation can be pointwise verified.) We define the concept of weak solutions, requiring no more diffenrentiability than that they belong to the first order Sobolev space $W^{1,p}(\Omega)$. Even the local space $W^{1,p}_{\text{loc}}(\Omega)$ will do.

Definition 2.5 Let Ω be a domain in $\mathbb{R}^n$. We say that $u \in W^{1,p}_{\text{loc}}(\Omega)$ is a *weak solution* of the *p*-Laplace equation in Ω, if

$$\int \langle |\nabla u|^{p-2} \nabla u, \nabla \eta \rangle dx = 0 \qquad (2.6)$$

for each $\eta \in C_0^\infty(\Omega)$. If, in addition, u is continuous, then we say that u is a *p-harmonic function*.

We naturally read $|0|^{p-2}0$ as 0 also when $1 < p < 2$. As we will see in section 3, all weak solutions are continuous. In fact, every weak solution can be redefined in a set of zero Lebesgue measure so that the new function is continuous. When appropriate, we assume that the redefinition has been performed.

We have the following basic result.

Theorem 2.7 *The following conditions are equivalent for $u \in W^{1,p}(\Omega)$:*

(i) u is minimizing:

$$\int |\nabla u|^p dx \le \int |\nabla v|^p dx, \quad \text{when } v - u \in W_0^{1,p}(\Omega).$$

(ii) the first variation vanishes:

$$\int \langle |\nabla u|^{p-2} \nabla u, \nabla \eta \rangle dx = 0, \quad \text{when } \eta \in W_0^{1,p}(\Omega).$$

If, in addition, $\Delta_p u$ is continuous, then the conditions are equivalent to the pointwise equation $\Delta_p u = 0$ in Ω.

Remark If 2.6 holds for all $\eta \in C_0^\infty(\Omega)$, then it also holds for all $\eta \in W_0^{1,p}(\Omega)$, if we know that $u \in W^{1,p}(\Omega)$. Thus the minimizers are the same as the weak solutions.

Proof "(i) $\Rightarrow$ (ii)". We use a device due to Lagrange. If u is minimizing, select

$$v(x) = u(x) + \varepsilon \eta(x),$$

where ε is a real parameter. Since

$$J(\varepsilon) = \int_\Omega |\nabla(u + \varepsilon \eta)|^p dx$$

attains its minimum for $\varepsilon = 0$, we must have $J'(0) = 0$ by the infinitesimal calculus. This is (ii).

"(ii) $\Rightarrow$ (i)" The inequality

$$|b|^p \geq |a|^p + p\langle|a|^{p-2}a, b - a\rangle$$

holds for vectors (if $p \geq 1$) by convexity. It follows that

$$\int_\Omega |\nabla v|^p dx \geq \int_\Omega |\nabla u|^p dx + p \int_\Omega \langle|\nabla u|^{p-2}\nabla u, \nabla(v - u)\rangle dx.$$

If (ii) is valid, take $\eta = v - u$ to see that the last integral vanishes. This is (i).
    Finally, the equivalence of (ii) and the extra condition is obtained from (2.3). $\square$

Before proceeding, we remark that the operator

$$\Delta_p u = |\nabla u|^{p-4}\left\{|\nabla u|^2 \Delta u + (p - 2)\sum \frac{\partial u}{\partial x_i}\frac{\partial u}{\partial x_j}\frac{\partial^2 u}{\partial x_i \partial x_j}\right\}$$

is not well defined at points where $\nabla u = 0$ in the case $1 < p < 2$, at least not for
arbitrary smooth functions. In the case $p \geq 2$ one can divide out the crucial factor.
Actually, the weak solutions $u \in C^2(\Omega)$ are precisely characterized by the equation

$$|\nabla u|^2 \Delta u + (p - 2)\sum \frac{\partial u}{\partial x_i}\frac{\partial u}{\partial x_j}\frac{\partial^2 u}{\partial x_i \partial x_j} = 0 \qquad (2.8)$$

for all p in the range $1 < p < \infty$. The proof for $p < 2$ is difficult, cf [JLM]. The
reader may think of the simpler problem: Why are the equations $|\nabla u|\Delta u = 0$ and
$\Delta u = 0$ equivalent for $u \in C^2$?
    Let us return to Definition 2.5 and derive some preliminary estimates from the
weak form of the equation. The art is to find the right test function. We will often
use the notation

$$B_r = B(x_0, r), \quad B_{2r} = B(x_0, 2r)$$

for *concentric* balls of radii r and $2r$, respectively.

Lemma 2.9 (Caccioppoli) *If u is a weak solution in Ω, then*

$$\int_\Omega \zeta^p |\nabla u|^p dx \leq p^p \int_\Omega |u|^p |\nabla \zeta|^p dx \qquad (2.10)$$

for each $\zeta \in C_0^\infty(\Omega)$, $0 \leq \zeta \leq 1$. In particular, if $B_{2r} \subset \Omega$, then

$$\int_{B_r} |\nabla u|^p dx \leq p^p r^{-p} \int_{B_{2r}} |u|^p dx. \qquad (2.11)$$

Proof Use

$$\eta = \zeta^p u,$$
$$\nabla\eta = \zeta^p \nabla u + p\zeta^{p-1} u \nabla\zeta.$$

By the Eq. (2.6) and Hölder's inequality

$$\int_\Omega \zeta^p |\nabla u|^p dx = -p \int_\Omega \zeta^{p-1} u \langle |\nabla u|^{p-2} \nabla u, \nabla\zeta \rangle dx$$

$$\leq p \int_\Omega |\zeta \nabla u|^{p-1} |u \nabla\zeta| dx$$

$$\leq p \left\{ \int_\Omega \zeta^p |\nabla u|^p dx \right\}^{1-\frac{1}{p}} \left\{ \int_\Omega |u|^p |\nabla\zeta|^p dx \right\}^{\frac{1}{p}}.$$

The estimate follows.

Finally, if $B_{2r} \subset \Omega$, we may choose ζ as a radial function satisfying $\zeta = 1$ in B_r, $|\nabla\zeta| \leq r^{-1}$ and $\zeta = 0$ outside B_{2r}. This is possible by approximation. This concludes the proof.   □

Occasionally, it is useful to consider weak supersolutions and weak subsolutions. As a mnemonic rule, "$\Delta_p v \leq 0$" for supersolutions and "$\Delta_p u \geq 0$" for subsolutions.

Definition 2.12 We say that $v \in W^{1,p}_{loc}(\Omega)$ is a *weak supersolution* in Ω, if

$$\int_\Omega \langle |\nabla v|^{p-2} \nabla v, \nabla\eta \rangle dx \geq 0 \tag{2.13}$$

for all nonnegative $\eta \in C_0^\infty(\Omega)$. For *weak subsolutions* the inequality is reversed.

In the a priori estimate below it is remarkable that the majorant is independent of the weak supersolution itself.

Lemma 2.14 *If $v > 0$ is a weak supersolution in Ω, then*

$$\int_\Omega \zeta^p |\nabla \log v|^p dx \leq \left(\frac{p}{p-1} \right)^p \int_\Omega |\nabla\zeta|^p dx$$

whenever $\zeta \in C_0^\infty(\zeta)$, $\zeta \geq 0$.

Proof One may add constants to the weak supersolutions. First, prove the estimate for $v(x) + \varepsilon$ in place of $v(x)$. Then let $\varepsilon \to 0$ in

$$\int_\Omega \frac{\zeta^p |\nabla v|^p}{(v+\varepsilon)^p} dx \leq \left(\frac{p}{p-1} \right)^p \int_\Omega |\nabla\zeta|^p dx.$$

Hence we may assume that $v(x) \geq \varepsilon > 0$. Next use the test function $\eta = \zeta^p v^{1-p}$. Then

$$\nabla\eta = p\zeta^{p-1}v^{1-p}\nabla\zeta - (p-1)\zeta^p v^{-p}\nabla v$$

and we obtain

$$(p-1)\int_\Omega \zeta^p v^{-p}|\nabla v|^p dx \leq p\int_\Omega \zeta^{p-1}v^{1-p}\langle|\nabla v|^{p-2}\nabla v, \nabla\zeta\rangle dx$$

$$\leq p\int_\Omega \zeta^{p-1}v^{1-p}|\nabla v|^{p-1}|\nabla\zeta| dx$$

$$\leq p\left\{\int_\Omega \zeta^p v^{-p}|\nabla v|^p dx\right\}^{1-\frac{1}{p}}\left\{\int_\Omega |\nabla\zeta|^p dx\right\}^{\frac{1}{p}},$$

from which the result follows. □

The Comparison Principle, which in the linear case is merely a restatement of the Maximum Principle, is one of the cornerstones in the theory.

Theorem 2.15 (Comparison Principle) *Suppose that u and v are p-harmonic functions in a bounded domain Ω. If at each $\zeta \in \partial\Omega$*

$$\limsup_{x\to\zeta} u(x) \leq \liminf_{x\to\zeta} v(x),$$

excluding the situation $\infty \leq \infty$ and $-\infty \leq -\infty$, then $u \leq v$ in Ω.

Proof Given $\varepsilon > 0$, the open set

$$D_\varepsilon = \{x\,|\,u(x) > v(x) + \varepsilon\}$$

is empty or $D_\varepsilon \subset\subset \Omega$. Subtracting the equations we get

$$\int_\Omega \langle|\nabla v|^{p-2}\nabla v - |\nabla u|^{p-2}\nabla u, \nabla\eta\rangle dx = 0$$

for all $\eta \in W_0^{1,p}(\Omega)$ with compact support in Ω. The choice

$$\eta(x) = \min\{v(x) - u(x) + \varepsilon, 0\}$$

yields

$$\int_{D_\varepsilon} \langle|\nabla v|^{p-2}\nabla v - |\nabla u|^{p-2}\nabla u, \nabla v - \nabla u\rangle dx = 0.$$

This is possible only if $\nabla u = \nabla v$ a.e. in D_ε, because the integrand is positive when $\nabla u \neq \nabla v$. Thus $u(x) = v(x) + C$ in D_ε and $C = \varepsilon$ because $u(x) = v(x) + \varepsilon$ on ∂D_ε. Thus $u \leq v + \varepsilon$ in Ω. It follows that $u \leq v$.   □

Remark The Comparison Principle also holds when u is a weak subsolution and v a weak supersolution. Then $u \leq v$ is valid a.e. in Ω.

   The next topic is the *existence* of a p-harmonic function with given boundary values. One can use the Lax-Milgram theorem, but I prefer the direct method in the Calculus of Variations, due to Lebesgue in 1907. The starting point is the variational integral (2.1), the Dirichlet integral with p.

Theorem 2.16 *Suppose that $g \in W^{1,p}(\Omega)$, where Ω is a bounded domain in $\mathbf{R}^n$. There exists a unique $u \in W^{1,p}(\Omega)$ with the boundary values $u - g \in W_0^{1,p}(\Omega)$ such that*

$$\int_\Omega |\nabla u|^p \, dx \leq \int_\Omega |\nabla v|^p \, dx$$

for all similar v. Thus u is a weak solution. In fact, $u \in C(\Omega)$ after a redefinition. If, in addition, $g \in C(\overline{\Omega})$ and if the boundary $\partial \Omega$ is regular enough, then $u \in C(\overline{\Omega})$ and $u|_{\partial \Omega} = g|_{\partial \Omega}$.

Proof Let us begin with the uniqueness, which is a consequence of strict convexity. If there were two minimizers, say u_1 and u_2, we could choose the competing function $v = (u_1 + u_2)/2$ and use

$$\left| \frac{\nabla u_1 + \nabla u_2}{2} \right|^p \leq \frac{|\nabla u_1|^p + |\nabla u_2|^p}{2}.$$

If $\nabla u_1 \neq \nabla u_2$ in a set of positive measure, then the above inequality is strict there and it would follow that

$$\int_\Omega |\nabla u_2|^p \, dx \leq \int_\Omega \left| \frac{\nabla u_1 + \nabla u_2}{2} \right|^p \, dx < \int_\Omega \frac{|\nabla u_1|^p + |\nabla u_2|^p}{2} \, dx$$

$$= \frac{1}{2} \int_\Omega |\nabla u_1|^p \, dx + \frac{1}{2} \int_\Omega |\nabla u_2|^p \, dx = \int_\Omega |\nabla u_2|^p \, dx,$$

which is a clear contradicition. Thus $\nabla u_1 = \nabla u_2$ a.e. in Ω and hence $u_1 = u_2 +$ Constant. The constant of integration is zero, because $u_2 - u_1 \in W_0^{1,p}(\Omega)$. This proves the uniqueness.
   The existence of a minimizer is obtained through the so-called direct method, see [D] and [G]. Let

$$I_0 = \inf \int_\Omega |\nabla v|^p \, dx \leq \int_\Omega |\nabla g|^p \, dx < \infty.$$

Thus $0 \leq I_0 < \infty$. Choose admissible functions v_j such that

$$\int_\Omega |\nabla v_j|^p dx < I_0 + \frac{1}{j}, \quad j = 1, 2, 3, \ldots \tag{2.17}$$

We aim at bounding the sequence $\|v_j\|_{W^{1,p}(\Omega)}$. The inequality

$$\|w\|_{L^p(\Omega)} \leq C_\Omega \|\nabla w\|_{L^p(\Omega)}$$

holds for all $w \in W_0^{1,p}(\Omega)$, and in particular for $w = v_j - g$. We obtain

$$\|v_j - g\|_{L^p(\Omega)} \leq C_\Omega \{\|\nabla v_j\|_{L^p(\Omega)} + \|\nabla g\|_{L^p(\Omega)}\}$$
$$\leq C_\Omega \{(I_0 + 1)^{\frac{1}{p}} + \|\nabla g\|_{L^p(\Omega)}\}$$

Now it follows from the triangle inequality that

$$\|v_j\|_{L^p(\Omega)} \leq M \quad (j = 1, 2, 3, \ldots) \tag{2.18}$$

where the constant M is independent of the index j. Together Eqs. 2.17 and 2.18 constitute the desired bound.

By weak compactness there exist a function $u \in W^{1,p}(\Omega)$ and a subsequence such that

$$v_{j_\nu} \rightharpoonup u, \ \nabla v_{j_\nu} \rightharpoonup \nabla u \ \text{weakly in } L^p(\Omega).$$

We have $u - g \in W_0^{1,p}(\Omega)$, because this space is closed under weak convergence. Thus u is an admissible function. We claim that u is also the minimizer sought for. By weak lower semicontinuity

$$\int_\Omega |\nabla u|^p dx \leq \lim_{\nu \to \infty} \int_\Omega |\nabla v_{j_\nu}|^p dx = I_0$$

and the claim follows. (This can also be deduced from

$$\int_\Omega |\nabla v_{j_\nu}|^p dx \geq \int_\Omega |\nabla u|^p dx + p \int_\Omega \langle |\nabla u|^{p-2} \nabla u, \nabla v_{j_\nu} - \nabla u \rangle dx$$

since the last integral approaches zero. Recall that

$$|b|^p \geq |a|^p + p\langle |a|^{p-2} a, b - a \rangle, \quad p \geq 1,$$

holds for vectors.) We remark that, a posteriori, one can verify that the minimizing sequence converges strongly in the Sobolev norm.

For the rest of the proof we mention that the continuity will be treated in section 3 and the question about classical boundary values is postponed till section 6. □

A retrospect of the previous proof of existence reveals that we have avoided some dangerous pitfalls. First, if we merely assume that the boundary values are continuous, say $g \in C(\overline{\Omega})$, it may so happen that $I(v) = \infty$ for each reasonable function $v \in C(\overline{\Omega})$ with these boundary values g. Indeed, J. Hadamard has given such an example for $p = n = 2$. If we take Ω as the unit disc in the plane and define

$$g(r, \theta) = \sum_{j=1}^{\infty} \frac{r^{j!} \cos(j!\theta)}{j^2}$$

in polar coordinates, we have the example. The function $g(r, \theta)$ is harmonic when $r < 1$ and continuous when $r \leq 1$ (use Weierstrass's test for uniform convergence). The Dirichlet integral of g is infinite. –Notice that we have avoided the phenomenon, encountered by Hadamard, by assuming that g belongs to a Sobolev space.

The second remark is a celebrated example of Weierstrass. He observed that the one-dimensional variational integral

$$I(u) = \int_{-1}^{1} x^2 u'(x)^2 dx$$

has no continuous minimizer with the "boundary values" $u(-1) = -1$ and $u(+1) = +1$. The weight function x^2 is catastrophical near the origin. The example can be generalized. This indicates that some care is called for, when it comes to questions about existence.

We find it appropriate to give a quantitative formulation of the continuity of the weak solutions, although the proof is postponed.

Theorem 2.19 *Suppose that $u \in W_{loc}^{1,p}(\Omega)$ is a weak solution to the p-harmonic equation. Then*

$$|u(x) - u(y)| \leq L|x - y|^{\alpha}$$

for a.e. $x, y \in B(x_0, r)$ provided that $B(x_0, 2r) \subset\subset \Omega$. The exponent $\alpha > 0$ depends only on n and p, while L also depends on $\|u\|_{L^p(B_{2r})}$.

We shall deduce the theorem from the so-called Harnack inequality, given below and proved in section 3. We write $B_r = B(x_0, r)$.

Theorem 2.20 (Harnack's inequality) *Suppose that $u \in W_{loc}^{1,p}(\Omega)$ is a weak solution and that $u \geq 0$ in $B_{2r} \subset \Omega$. Then the quantities*

$$m(r) = \operatorname*{ess\,inf}_{B_r} u, \quad M(r) = \operatorname*{ess\,sup}_{B_r} u$$

satisfy

$$M(r) \le Cm(r)$$

where $C = C(n, p)$.

The main feature is that *the same constant C will do for all weak solutions.*

Since one may add constants to solutions, *the Harnack inequality implies Hölder continuity.* To see this, first apply the inequality to the two non-negative weak solutions $u(x) - m(2r)$ and $M(2r) - u(x)$, where r is small enough. It follows that

$$M(r) - m(2r) \le C(m(r) - m(2r)),$$
$$M(2r) - m(r) \le C(M(2r) - M(r)).$$

Hence

$$\omega(r) \le \frac{C - 1}{C + 1}\omega(2r)$$

where $\omega(r) = M(r) - m(r)$ is the (essential) oscillation of u over $B(x_0, r)$. It is decisive that

$$\lambda = \frac{C - 1}{C + 1} < 1.$$

Iterating $\omega(r) \le \lambda\omega(2r)$, we get $\omega(2^{-k}r) \le \lambda^k\omega(r)$. We conclude that

$$\omega(\varrho) \le A\left(\frac{\varrho}{r}\right)^{\alpha}\omega(r), \quad 0 < \rho < r$$

for some $\alpha = \alpha(n, p) > 0$ and $A = A(n, p)$.

Thus we have proved that Harnack's inequality implies Hölder continuity,[1] provided that we already know that also sign changing solutions are locally bounded. The possibility $\omega(r) = \infty$ is eliminated in Corollary 3.8. If the domain of definition is the whole space, we get Liouville's Theorem.

Corollary 2.21 (Liouville) *If a p-harmonic function is bounded from below (or from above) in the entire space $\mathbf{R}^n$, then it reduces to a constant.*

Finally, we point out a simple but important property, the Strong Maximum Principle.

Corollary 2.22 (Strong Maximum Principle) *If a p-harmonic function attains its maximum at an interior point, then it reduces to a constant.*

Proof If $u(x_0) = \max_{x \in \Omega} u(x)$ for $x_0 \in \Omega$, then we can apply the Harnack inequality on the p-harmonic function $u(x_0) - u(x)$, which indeed is non-negative. It follows that $u(x) = u(x_0)$, when $2|x - x_0| < \mathrm{dist}(x_0, \partial\Omega)$. Through a chain of intersecting balls the identity $u(x) = u(x_0)$ is achieved at an arbitrary point x in Ω.   □

---

[1] "Was it Plato who made his arguments by telling a story with an obvious flaw, and allowing the listener to realize the error?"

Remark Of course, also the corresponding Strong Minimum Principle holds. However, a strong version of the Comparison Principle is not known in several dimensions, $n \geq 3$, when $p \neq 2$.

Chapter 3
Regularity Theory

The weak solutions of the p-harmonic equation are, by definition, members of the Sobolev space $W^{1,p}_{loc}(\Omega)$. In fact, they are also of class $C^{\alpha}_{loc}(\Omega)$. More precisely, a weak solution can be redefined in a set of Lebesgue measure zero, so that the new function is locally Hölder continuous with exponent $\alpha = \alpha(n, p)$. Actually, a deeper and stronger regularity result holds. In 1968 N.Ural'tseva proved that even the gradient is locally Hölder continuous; we refer to [Ur, Db1, E, Uh, Le2, To] for this $C^{1,\alpha}_{loc}$ result.

To obtain the Hölder continuity of the weak solutions one had better distinguish between three cases, depending on the value of p. Recall that n is the dimension.

(1) If $p > n$, then every function in $W^{1,p}(\Omega)$ is continuous.
(2) The case $p = n$ (the so-called borderline case) is rather simple, but requires a proof. We will present a proof based on "the hole filling technique" of Widman.
(3) The case $p < n$ is much harder. Here the regularity theory of elliptic equations is called for. There are essentially three methods, developed by

- E. De Giorgi 1957
- J. Nash 1958
- J. Moser 1961

to prove the Hölder continuity in a wide class of partial differential equations. While De Georgi's method is the most robust, we will, nevertheless, use Moser's approach, which is very elegant. Thus we will present the so-called Moser iteration, which leads to Harnack's inequality. A short presentation for $p = 2$ can be found in [J]. See also [Mo2]. The general p is in [T1]. De Giorgi's method is in [Dg, G, LU]. For an alternative proof of the case $p > n - 2$ and $p \geq 2$ see the remark after the proof of Theorem 4.1.

© The Author(s), under exclusive license to Springer Nature Switzerland AG 2019
P. Lindqvist, *Notes on the Stationary p-Laplace Equation*,
SpringerBriefs in Mathematics, https://doi.org/10.1007/978-3-030-14501-9_3

3.1 The Case $p > n$

In this case all functions in the Sobolev space $W^{1,p}_{\mathrm{loc}}(\Omega)$ are continuous. Indeed, if $p > n$ and $v \in W^{1,p}(B)$ where B is a ball (or a cube) in $\mathbf{R}^n$, then

$$|v(y) - v(x)| \le C_1 |x - y|^{1 - \frac{n}{p}} \|\nabla v\|_{L^p(B)} \tag{3.1}$$

when $x, y \in B$, cf [GT, Theorem 7.17]. The Hölder exponent is $\alpha = 1 - \frac{n}{p}$. If u is a positive weak solution or supersolution, Lemma 2.14 implies

$$\|\nabla \log u\|_{L^p(B_r)} \le C_2 r^{\frac{n-p}{p}} \tag{3.2}$$

assuming that $u > 0$ in B_{2r}. For $v = \log u$ we obtain

$$\left| \log \frac{u(y)}{u(x)} \right| \le C_1 C_2. \tag{3.3}$$

This is Harnack's inequality (see Theorem 2.20) with the constant $C(n, p) = e^{C_1 C_2}$.

In the favourable case $p > n$ a remarkable property holds for the Dirichlet problem: all *the boundary points of an arbitrary domain are regular.* Indeed, if Ω is a bounded domain in $\mathbf{R}^n$ and if $g \in C(\overline{\Omega}) \cap W^{1,p}(\Omega)$ is given, there exists a p-harmonic function $u \in C(\overline{\Omega}) \cap W^{1,p}(\Omega)$ such that $u = g$ on $\partial\Omega$. The boundary values are attained, not only in Sobolev's sense, but also in the classical sense. This follows from the general inequality

$$|v(y) - v(x)| \le C_\Omega |x - y|^{1 - \frac{n}{p}} \|\nabla v\|_{L^p(\Omega)}$$

valid for all $v \in W^{1,p}_0(\Omega)$. Hence $v \in C^\alpha(\overline{\Omega})$ and $v = 0$ on $\partial\Omega$. The argument is to apply the inequality to a minimizing sequence. See also Sect. 6.

3.2 The Case $p = n$

The proof of the Hölder continuity is based on the so-called hole filling technique (due to Widman, see [Wi]) and the following elementary lemma. We do not seem to reach Harnack's inequality this way.

Lemma 3.4 (Morrey) *Assume that* $u \in W^{1,p}(\Omega)$, $1 \le p < \infty$. *Suppose that*

$$\int_{B_r} |\nabla u|^p dx \le K r^{n - p + p\alpha} \tag{3.5}$$

whenever $B_{2r} \subset \Omega$. *Here $0 < \alpha \leq 1$ and K are independent of the ball B_r. Then $u \in C_{loc}^{\alpha}(\Omega)$. In fact,*

$$\underset{B_r}{osc}(u) \leq \frac{4}{\alpha} \left(\frac{K}{\omega_n} \right)^{\frac{1}{p}} r^{\alpha}, \quad B_{2r} \subset \Omega.$$

Proof See [LU, Chap. 2, Lemma 4.1, p. 56] or [GT, Theorem 7.19].

For the continuity proof, we let $B_{2r} = B(x_0, 2r) \subset\subset \Omega$. Select a radial test function ζ such that $0 \leq \zeta \leq 1$, $\zeta = 1$ in B_r, $\zeta = 0$ outside B_{2r} and $|\nabla \zeta| \leq r^{-1}$. Choose

$$\eta(x) = \zeta(x)^n (u(x) - a)$$

in the n-harmonic equation. This yields

$$\int_{\Omega} \zeta^n |\nabla u|^n dx = -n \int_{\Omega} \zeta^{n-1} (u - a) \langle |\nabla u|^{n-2} \nabla u, \nabla \zeta \rangle dx$$

$$\leq n \int_{\Omega} |\zeta \nabla u|^{n-1} |(u - a) \nabla \zeta| dx$$

$$\leq n \left\{ \int_{\Omega} \zeta^n |\nabla u|^n dx \right\}^{1 - \frac{1}{n}} \left\{ \int_{\Omega} |u - a|^n |\nabla \zeta|^n dx \right\}^{\frac{1}{n}}.$$

It follows that

$$\int_{B_r} |\nabla u|^n dx \leq n^n r^{-n} \int_{B_{2r} \setminus B_r} |u - a|^n dx.$$

The constant a is at our disposal. Let a denote the average

$$a = \frac{1}{|H(r)|} \int_{H(r)} u(x) dx$$

of u taken over the annulus $H(r) = B_{2r} \setminus B_r$. The Poincaré inequality

$$\int_{H(r)} |u(x) - a|^n dx \leq C r^n \int_{H(r)} |\nabla u|^n dx$$

yields

$$\int_{B_r} |\nabla u|^n dx \leq C n^n \int_{H(r)} |\nabla u|^n dx.$$

Now the trick comes. Add $Cn^n \int_{B_r} |\nabla u|^n dx$ to both sides of the last inequality. This fills the hole in the annulus and we obtain

$$(1 + Cn^n) \int_{B_r} |\nabla u|^n dx \leq Cn^n \int_{B_{2r}} |\nabla u|^n dx.$$

In other words

$$D(r) \leq \lambda D(2r), \quad \lambda < 1,$$

holds for the Dirichlet integral

$$D(r) = \int_{B_r} |\nabla u|^n dx$$

with the constant

$$\lambda = \frac{Cn^n}{1 + Cn^n} < 1.$$

By iteration

$$D(2^{-k}r) \leq \lambda^k D(r), \quad k = 1, 2, 3, \ldots$$

A calculation reveals that

$$D(\varrho) \leq 2^\delta (\frac{\varrho}{r})^\delta D(r), \quad 0 < \varrho < r,$$

with $\delta = \log(1/\lambda) : \log 2$, when $B_{2r} \subset \Omega$. This is the estimate called for in Morrey's lemma. The Hölder continuity follows.   □

Remark A careful analysis of the above proof shows that it works for all p in a small range $(n - \varepsilon, n]$, where $\varepsilon = \varepsilon(n, p)$.

3.3 The Case $1 < p < n$

This is much more difficult than the case $p \geq n$. The idea of Moser's proof is to reach the Harnack inequality

$$\operatorname*{ess\,sup}_{B} u \leq C \operatorname*{ess\,inf}_{B} u$$

through the limits

$$\operatorname*{ess\,sup}_{B} u = \lim_{q \to \infty} \left\{ \int_B u^q dx \right\}^{\frac{1}{q}}$$

$$\operatorname*{ess\,inf}_{B} u = \lim_{q \to -\infty} \left\{ \int_B u^q dx \right\}^{\frac{1}{q}}$$

The equation is used to deduce reverse Hölder inequalities like

$$\left\{ \int_{B_r} u^{p_2} dx \right\}^{\frac{1}{p_2}} \le K \left\{ \int_{B_R} u^{p_1} dx \right\}^{\frac{1}{p_1}}$$

where $-\infty < p_1 < p_2 < \infty$ and $0 < r < R$. The "constant" K will typically blow up as $r \to R$, and, since one does not reach all exponents at one stroke, one has to pay attention to this, when using the reverse Hölder inequality infinitely many times.

Several lemmas are needed and it is convenient to include weak subsolutions and supersolutions. In the first lemma we do not assume positivity, because we need it to conclude that arbitrary solutions are locally bounded.

Lemma 3.6 *Let $u \in W^{1,p}_{loc}(\Omega)$ be a weak subsolution. Then*

$$\operatorname*{ess\,sup}_{B_r}(u_+) \le C_\beta \left\{ \frac{1}{(R-r)^n} \int_{B_R} u_+^\beta dx \right\}^{\frac{1}{\beta}} \tag{3.7}$$

for $\beta > p - 1$ when $B_R \subset\subset \Omega$. Here $u_+ = \max\{u(x), 0\}$ and $C_\beta = C(n, p, \beta)$.

Proof The proof has two major steps. First, the test function $\eta = \zeta^p u_+^{\beta-(p-1)}$ is used to produce the estimate

$$\left\{ \int_{B_r} u_+^{\kappa\beta} dx \right\}^{\frac{1}{\kappa\beta}} \le C^{\frac{1}{\beta}} \left(\frac{2\beta - p + 1}{\beta - p + 1} \right)^{\frac{p}{\beta}} \frac{1}{(R-r)^{\frac{p}{\beta}}} \left\{ \int_{B_R} u_+^\beta dx \right\}^{\frac{1}{\beta}}$$

where $\kappa = n/(n-p)$ and $\beta > p - 1$. Second, the above estimate is iterated so that the exponents $\kappa\beta$, $\kappa^2\beta$, $\kappa^3\beta$, ... are reached, while the radii shrink.

Write $\alpha = \beta - (p-1) > 0$. We insert

$$\nabla\eta = p\zeta^{p-1} u_+^\alpha \nabla\zeta + \alpha u_+^{\alpha-1} \zeta^p \nabla u_+$$

into the equation. This yields

$$\alpha \int_\Omega \zeta^p u_+^{\alpha-1} |\nabla u_+|^p dx = -p \int_\Omega \zeta^{p-1} u_+^\alpha \langle |\nabla u_+|^{p-2}\nabla u_+, \nabla\zeta\rangle dx$$

since $\nabla u_+ = \nabla u$ a.e. in the set where $u \ge 0$.

For simplicity we write u instead of u_+ from now on. Use the decomposition

$$\alpha = \frac{(\alpha - 1)(p - 1)}{p} + \frac{\alpha + p - 1}{p}$$

to factorize u^α in Hölder's inequality. We obtain

$$\alpha \int_\Omega \zeta^p u^{\alpha-1} |\nabla u|^p dx$$

$$\leq p \int_\Omega \zeta^{p-1} u^{(\alpha-1)(p-1)/p} |\nabla u|^{p-1} \cdot u^{\beta/p} |\nabla \zeta| dx$$

$$\leq p \Big\{ \int_\Omega \zeta^p u^{\alpha-1} |\nabla u|^p dx \Big\}^{1 - \frac{1}{p}} \Big\{ \int_\Omega u^\beta |\nabla \zeta|^p dx \Big\}^{\frac{1}{p}}.$$

Divide out the common factor (an integral) and rise everything to the pth power. We arrive at

$$\int_\Omega \zeta^p u^{\alpha-1} |\nabla u|^p dx \leq \Big(\frac{p}{\alpha}\Big)^p \int_\Omega u^\beta |\nabla \zeta|^p dx,$$

which can be written as

$$\int_\Omega |\zeta \nabla u^{\beta/p}|^p dx \leq \Big(\frac{\beta}{\beta - (p - 1)}\Big)^p \int_\Omega |u^{\beta/p} \nabla \zeta|^p dx.$$

Use

$$|\nabla(\zeta u^{\beta/p})| \leq |\zeta \nabla u^{\beta/p}| + |u^{\beta/p} \nabla \zeta|$$

and Minkowski's inequality to obtain

$$\int_\Omega |\nabla(\zeta u^{\beta/p})|^p dx \leq \Big(\frac{2\beta - p + 1}{\beta - p + 1}\Big)^p \int_\Omega |u^{\beta/p} \nabla \zeta|^p dx.$$

According to Sobolev's inequality (for the function $\zeta u^{\frac{\beta}{p}}$) we have

$$\Big\{ \int_\Omega |\zeta u^{\beta/p}|^{\kappa p} dx \Big\}^{\frac{1}{\kappa}} \leq S^p \int_\Omega |\nabla(\zeta u^{\beta/p})|^p dx$$

where $S = S(n, p)$. Recall, that, as usual $|\nabla \zeta| \leq 1/(R - r)$ and $\zeta = 1$ in B_r. It follows that

$$\left\{ \int_{B_r} u^{\kappa\beta} dx \right\}^{\frac{1}{\kappa\beta}} \leq \left\{ \left(S \frac{2\beta - p + 1}{\beta - p + 1} \frac{1}{R - r} \right)^p \int_{B_R} u^{\beta} dx \right\}^{\frac{1}{\beta}}.$$

We have accomplished the first step, a reverse Hölder inequality.

Next, let us iterate the estimate. Fix a β, say $\beta_0 > p - 1$ and notice that

$$\frac{2\beta - p + 1}{\beta - p + 1} \leq \frac{2\beta_0 - p + 1}{\beta_0 - p + 1} = b$$

when $\beta \geq \beta_0$. Start with β_0 and the radii $r_0 = R$ and $r_1 = r + (R - r)/2$ in the place of R and r. This yields

$$\|u\|_{\kappa\beta_0, r_1} \leq (Sb)^{p/\beta_0} \left(\frac{2}{R - r} \right)^{\frac{p}{\beta_0}} \|u\|_{\beta_0, r_0}$$

with the notation

$$\|u\|_{q, \varrho} = \left\{ \int_{B_\varrho} u^q dx \right\}^{\frac{1}{q}}.$$

Then use r_1 and $r_2 = r + 2^{-2}(R - r)$ to improve $\kappa\beta_0$ to $\kappa^2\beta_0$. Hence

$$\|u\|_{\kappa^2\beta_0, r_2} \leq (Sb)^{\frac{p}{\kappa\beta_0}} \left(\frac{4}{R - r} \right)^{\frac{p}{\kappa\beta_0}} \|u\|_{\kappa\beta_0, r_1}$$

$$\leq (Sb)^{\frac{p}{\beta_0} + \frac{p}{\kappa\beta_0}} \frac{2^{\frac{p}{\beta_0} + \frac{2p}{\kappa\beta_0}}}{(R - r)^{\frac{p}{\beta_0} + \frac{p}{\kappa\beta_0}}} \|u\|_{\beta_0, r_0}$$

Here we can discern a pattern. Continuing like this, using radii $r_j = r + 2^{-j}(R - r)$, we arrive at

$$\|u\|_{\kappa^{j+1}\beta_0, r_{j+1}} \leq \left(\frac{Sb}{R - r} \right)^{p\beta_0^{-1} \sum \kappa^{-k}} 2^{p\beta_0^{-1} \sum (k+1)\kappa^{-k}} \|u\|_{\beta_0, r_0}$$

where the index k is summed over $0, 1, 2, \ldots, j$. The sums in the exponents are convergent and, for example,

$$\sum \kappa^{-k} = \frac{1 - \kappa^{-j-1}}{1 - \kappa^{-1}} \to \frac{n}{p}$$

as $j \to \infty$. To conclude the proof, use

$$\|u\|_{\kappa^{j+1}\beta_0, r} \leq \|u\|_{\kappa^{j+1}\beta_0, r_{j+1}}$$

and let $j \to \infty$. The majorant contains $(R - r)$ to the correct power n/β_0.   □

Corollary 3.8 *The weak solutions to the p-harmonic equation are locally bounded.*

Proof Let $\beta = p$ and apply the lemma to u and $-u$.   □

The next lemma is for supersolutions. It is decisive that one may take the exponent $\beta > p - 1$, which is possible since $\kappa > 1$. Hence one can combine with Lemma 3.6, because of the overlap.

Lemma 3.9 *Let* $v \in W^{1,p}_{loc}(\Omega)$ *be a non-negative weak supersolution. Then*

$$\left\{\frac{1}{(R-r)^n}\int_{B_r} v^\beta dx\right\}^{\frac{1}{\beta}} \leq C(\varepsilon, \beta)\left\{\frac{1}{(R-r)^n}\int_{B_R} v^\varepsilon dx\right\}^{\frac{1}{\varepsilon}}, \qquad (3.10)$$

when $0 < \varepsilon < \beta < \kappa(p-1) = n(p-1)/(n-p)$ *and* $B_R \subset\subset \Omega$.

Proof We may assume that $v(x) \geq \sigma > 0$. Otherwise, first prove the lemma for $v(x) + \sigma$ and let $\sigma \to 0$ at the end. Use the test function

$$\eta = \zeta^p v^{\beta - (p-1)}$$

This yields

$$\left\{\int_{B_r} v^{\kappa\beta} dx\right\}^{\frac{1}{\kappa\beta}} \leq C^{\frac{1}{\beta}}\left(\frac{p-1}{p-1-\beta}\right)^{\frac{p}{\beta}}\frac{1}{(R-r)^{p/\beta}}\left\{\int_{B_R} v^\beta dx\right\}^{\frac{1}{\beta}}$$

for $0 < \beta < p - 1$. Notice that we can reach an exponent $\kappa\beta > p - 1$. The calculations are similar to those in Lemma 3.6 and are omitted.

An iteration of the estimate leads to the desired result. The details are skipped.   □

In the next lemma the exponent $\beta < 0$.

Lemma 3.11 *Suppose that* $v \in W^{1,p}_{loc}(\Omega)$ *is a non-negative supersolution. Then*

$$\left\{\frac{1}{(R-r)^n}\int_{B_R} v^\beta dx\right\}^{\frac{1}{\beta}} \leq C \operatorname*{ess\,inf}_{B_r} v \qquad (3.12)$$

when $\beta < 0$ *and* $B_R \subset\subset \Omega$. *The constant* C *is of the form* $c(n, p)^{-1/\beta}$.

Proof Use the test function $\eta = \zeta^p v^{\beta-(p-1)}$ again, but now $\beta < 0$. First we arrive at

$$\int_\Omega |\nabla(\zeta v^{\beta/p})|^p dx \leq \left(\frac{p-1-2\beta}{p-1-\beta}\right)^p \int_\Omega v^\beta |\nabla\zeta|^p dx$$

after some calculations, similar to those in the proof of Lemma 3.6. The constant is less than 2^p. Using the Sobolev inequality we can write

$$\left\{ \int_\Omega \zeta^{\kappa p} v^{\kappa \beta} dx \right\}^{\frac{1}{\kappa}} \le (2S)^p \int_\Omega v^\beta |\nabla \zeta|^p dx$$

where $S = S(n, p)$ and $\kappa = n/(n-p)$. The estimate

$$\left\{ \int_{B_r} v^{\kappa \beta} dx \right\}^{\frac{1}{\kappa}} \le \left(\frac{2S}{R-r} \right)^p \int_{B_R} v^\beta dx$$

follows. An iteration of the estimate with the radii $r_0 = R$, $r_1 = r + 2^{-1}(R - r)$, $r_2 = r + 2^{-2}(R - r), \dots$ yields, via the exponents $\beta, \kappa\beta, \kappa^2\beta, \dots$

$$\left\{ \int_{B_{r_j}} v^{\kappa^j \beta} dx \right\}^{\kappa^{-j}} \le \left(\frac{2S}{R-r} \right)^{p \sum \kappa^{-k}} 2^{p \sum (k+1)\kappa^{-k}} \int_{B_R} v^\beta dx$$

where

$$\sum \kappa^{-k} = 1 + \kappa^{-1} + \dots + \kappa^{-(j-1)}.$$

As $j \to \infty$ we obtain

$$\operatorname*{ess\,sup}_{B_r}(v^\beta) \le \left(\frac{2S}{R-r} \right)^n 2^{\frac{n^2}{p}} \int_{B_R} v^\beta dx$$

Taking into account that $\beta < 0$ we have reached the desired estimate. $\qquad\square$

Combining the estimates achieved so far in the case $1 < p < n$, we have the following bounds for non-negative weak solutions:

$$\operatorname*{ess\,sup}_{B_r} u \le C_1(\varepsilon, n, p) \left\{ \frac{1}{(R-r)^n} \int_{B_R} u^\varepsilon dx \right\}^{\frac{1}{\varepsilon}},$$

$$\operatorname*{ess\,inf}_{B_r} u \ge C_2(\varepsilon, n, p) \left\{ \frac{1}{(R-r)^n} \int_{B_R} u^{-\varepsilon} dx \right\}^{-\frac{1}{\varepsilon}}$$

for all $\varepsilon > 0$. Take $R = 2r$. The missing link is the inequality

$$\left\{ \fint_{B_R} u^\varepsilon dx \right\}^{\frac{1}{\varepsilon}} \le C \left\{ \fint_{B_R} u^{-\varepsilon} dx \right\}^{-\frac{1}{\varepsilon}}$$

for some small $\varepsilon > 0$. The passage from negative to positive exponents is delicate. The gap in the iteration scheme can be bridged over with the help of the John-Nirenberg theorem, which is valid for functions in L^1. Its proof is in [JN] or [G, Sect. 2.4]. The weaker version given in [GT, Theorem 7.21] will do.

Theorem 3.13 (John-Nirenberg) *Let $w \in L^1_{loc}(\Omega)$. Suppose that there is a constant K such that*

$$\fint_{B_r} |w(x) - w_{B_r}| dx \leq K \tag{3.14}$$

holds whenever $B_{2r} \subset \Omega$. Then there exists a constant $\nu = \nu(n) > 0$ such that

$$\fint_{B_r} e^{\nu |w(x) - w_{B_r}|/K} dx \leq 2 \tag{3.15}$$

whenever $B_{2r} \subset \Omega$. (It also holds when $\overline{B_r} \subset \Omega$.)

The notation

$$w_{B_r} = \frac{\int\limits_{B_r} w(x) dx}{\int\limits_{B_r} dx} = \fint_{B_r} w \, dx$$

was used.

The two inequalities

$$\fint_{B_r} e^{\pm \nu (w(x) - w_{B_r})/K} dx \leq 2$$

follow immediately. Multiplying them we arrive at

$$\fint_{B_r} e^{\nu w(x)/K} dx \fint_{B_r} e^{-\nu w(x)/K} dx \leq 4 \tag{3.16}$$

since the constant factors $e^{\pm \nu w_{B_r}/K}$ cancel.

Next we use $w = \log u$ for the passage from negative to positive exponents. First we show that $w = \log u$ satisfies (3.14). Then we can conclude from (3.16) that

$$\fint_{B_r} u^{\nu/K} dx \cdot \fint_{B_r} u^{-\nu/K} dx \leq 4.$$

Writing $\varepsilon = \nu/K$ we have "the missing link"

$$\left\{ \fint_{B_r} u^\varepsilon dx \right\}^{\frac{1}{\varepsilon}} \leq 4^{\frac{1}{\varepsilon}} \left\{ \fint_{B_r} u^{-\varepsilon} dx \right\}^{-\frac{1}{\varepsilon}} \tag{3.17}$$

when $B_{2r} \subset\subset \Omega$.

To complete the first step, assume to begin with that $u > 0$ is a weak solution. Combining the Poincaré inequality

$$\int_{B_r} |\log u(x) - (\log u)_{B_r}|^p dx \le C_1 r^p \int_{B_r} |\nabla \log u|^p dx$$

with the estimate

$$\int_{B_r} |\nabla \log u|^p dx \le C_2 r^{n-p}$$

from lemma 2.14, we obtain for $B_{2r} \subset\subset \Omega$

$$\fint_{B_r} |w - w_{B_r}|^p dx \le C_1 C_2 \omega_n^{-1} = K.$$

This is the bound needed in the John-Nirenberg theorem. Finally, to replace $u > 0$ by $u \ge 0$, it is sufficient to observe that, if (3.17) holds for the weak solutions $u(x) + \sigma$, then it also holds for $u(x)$.

We have finished the proof of the Harnack inequality

$$M(r) \le Cm(r), \quad \text{when } B_{4r} \subset \Omega$$

in Theorem 2.20.

Remark It is possible to avoid the use of the John-Nirenberg inequality in the proof. To accomplish the zero passage one can use the equation in a more effective way by a more refined testing. Powers of $\log u$ appear in the test function and an extra iteration procedure is used. The original idea is in [BG]. See also [SC], [HL, Sect. 4.4, pp. 85-89] and [T2].

We record an inequality for weak supersolutions.

Corollary 3.18 *Suppose that $v \in W_{loc}^{1,p}(\Omega)$ is a non-negative supersolution. Then*

$$\left\{ \fint_{B_r} v^\beta dx \right\}^{\frac{1}{\beta}} \le C(n, p, \beta) \operatorname*{ess\,inf}_{B_r} v, \quad \beta < \frac{n(p-1)}{n-1}, \tag{3.19}$$

whenever $B_{2r} \subset \Omega$.

Proof This is a combination of (3.10), (3.12), and (3.17).   □

In fact, weak supersolutions are lower semicontinuos, after a possible redefinition in a set of measure zero. They are then pointwise defined.

Proposition 3.20 *A weak supersolution* $v \in W^{1,p}(\Omega)$ *is lower semicontinuous (after a redefinition in a set of measure zero). We can define*

$$v(x) = \operatorname{ess}\lim_{y \to x} \inf v(y)$$

pointwise. This representative is a lower semicontinuous function.[1]

Proof The case $p > n$ is clear, since then the Sobolev space contains only continuous functions (Morrey's inequality). In the range $p < n$ we claim that

$$v(x) = \operatorname{ess}\lim_{y \to x} \inf v(y)$$

at a.e. every x in Ω. The proof follows from this, since the right-hand side is always lower semicontinuous. For simplicity, we assume that v is locally bounded. Suppose that $p > 2n/(n+1)$ so that we may take $\beta = 1$ in the weak Harnack inequality (3.19). Use the function $v(x) - m(2r)$, where

$$m(r) = \operatorname*{ess\,inf}_{B_r} v.$$

We have

$$0 \leq \fint_{B_r} v \, dx - m(2r)$$

$$= \fint_{B_r} \left(v(x) - m(2r) \right) dx \leq C\left(m(r) - m(2r) \right).$$

Since $m(r)$ is monotone, $m(r) - m(2r) \to 0$ as $r \to 0$. It follows that

$$\operatorname{ess}\lim_{y \to x_0} \inf v(y) = \lim_{r \to 0} m(2r) = \lim_{r \to 0} \fint_{B(x_0,r)} v(x) \, dx$$

at *each* point x_0. Lebesgue's Differentiation Theorem states that the limit of the average on the right-hand side coincides with $v(x_0)$ at almost every point x_0.

If we are forced to take $\beta < 1$ in the weak Harnack inequality, a slight modification of the above proof will do. □

---

[1] It is a p-superharmonic function, see Definition 5.1.

Chapter 4
Differentiability

We have learned that the p-harmonic functions are Hölder continuous. In fact, much more regularity is valid. Even the gradients are locally Hölder continuous. In symbols, the function is of class $C^{1,\alpha}_{\text{loc}}(\Omega)$. More precisely, if u is p-harmonic in Ω and if $D \subset\subset \Omega$, then

$$|\nabla u(x) - \nabla u(y)|^{\alpha} \le L_D|x - y|$$

when $x, y \in D$. Here $\alpha = \alpha(n, p)$ and L_D depends on n, p, $\text{dist}(D, \partial\Omega)$ and $\|u\|_{\infty}$. This was proved in 1968 by N. Uraltseva, cf. [Ur]. We also refer to [E, Uh, Le2, Db1, To] about this difficult regularity question.[1] Here we are content with a weaker, but much simpler, result:

(1) If $1 < p \le 2$, then $u \in W^{2,p}_{\text{loc}}(\Omega)$; that means that u has second Sobolev derivatives.

(2) If $p \ge 2$, then $|\nabla u|^{(p-2)/2}\nabla u$ belongs to $W^{1,2}_{\text{loc}}(\Omega)$. Thus the Sobolev derivatives

$$\frac{\partial}{\partial x_j}\left(|\nabla u|^{\frac{p-2}{2}}\frac{\partial u}{\partial x_i}\right)$$

exist, but the passage to $\frac{\partial^2 u}{\partial x_i \partial x_j}$ is very difficult at the critical points ($\nabla u = 0$).

According to Lemma 5.1 on page 20 in [MW] the second derivatives exist when the *Cordes Condition*[2]

---

[1] The second Russian edition of the book [LU] by Ladyzhenskaya and Uraltseva includes the proof.

[2] For the equation

$$\sum a_{ij}(x, u, \nabla u)\frac{\partial^2 u}{\partial x_i \partial x_j} = a_0(x, u, \nabla u)$$

the Cordes Condition reads

$$\left(\sum_{j=1}^{n} a_{jj}\right)^2 \ge (n - 1 + \delta) \sum_{i,j=1}^{n} a_{ij}^2.$$

P. Lindqvist, *Notes on the Stationary p-Laplace Equation*,
SpringerBriefs in Mathematics, https://doi.org/10.1007/978-3-030-14501-9_4

$$1 < p < 3 + \frac{2}{n-2}$$

holds. Then $u \in W^{2,2}_{loc}(\Omega)$. To this one may add that u is real analytic (=is represented by the Taylor expansion) in the open set where $\nabla u \neq 0$, cf. [Le1, p.208].

We begin with the study of

$$F(x) = |\nabla u(x)|^{(p-2)/2} \nabla u(x)$$

in the case $p \geq 2$. It is plain that

$$\int_\Omega |F|^2 dx = \int_\Omega |\nabla u|^p dx.$$

Theorem 4.1 (Bojarski - Iwaniec) *Let $p \geq 2$. If u is p-harmonic in Ω, then $F \in W^{1,2}_{loc}(\Omega)$. For each subdomain $G \subset\subset \Omega$,*

$$\|DF\|_{L^2(G)} \leq \frac{C(n,p)}{\text{dist}(G, \partial\Omega)} \|F\|_{L^2(\Omega)}. \tag{4.2}$$

Proof The proof is taken from [BI1]. It is based on integrated difference quotients. Let $\zeta \in C_0^\infty(\Omega)$ be a cutoff function so that $0 \leq \zeta \leq 1$, $\zeta|G = 1$ and $|\nabla\zeta| \leq C_n/\text{dist}(G, \partial\Omega)$. (If required, replace Ω by a smaller domain Ω_1, $G \subset\subset \Omega_1 \subset\subset \Omega$.) We aim at difference quotients. Take $|h| < \text{dist}(\text{supp}\,\zeta, \partial\Omega)$. Notice that also $u_h = u(x+h)$ is p-harmonic, when $x + h \in \Omega$, h denoting a constant vector. The test function

$$\eta(x) = \zeta(x)^2(u(x+h) - u(x))$$

will do in the equations

$$\int_\Omega \langle |\nabla u(x)|^{p-2}\nabla u(x), \nabla\eta(x)\rangle dx = 0,$$

$$\int_\Omega \langle |\nabla u(x+h)|^{p-2}\nabla u(x+h), \nabla\eta(x)\rangle dx = 0.$$

Hence, after subtraction,

$$\int_\Omega \langle |\nabla u(x+h)|^{p-2}\nabla u(x+h) - |\nabla u(x)|^{p-2}\nabla u(x), \nabla\eta(x)\rangle dx = 0. \tag{4.3}$$

It follows that

$$\int_\Omega \zeta(x)\langle |\nabla u(x+h)|^{p-2}\nabla u(x+h) - |\nabla u(x)|^{p-2}\nabla u(x), \nabla u(x+h) - \nabla u(x)\rangle dx$$

$$= -2\int_\Omega \zeta(x)(u(x+h)-u(x))\langle |\nabla u(x+h)|^{p-2}\nabla u(x+h) - |\nabla u(x)|^{p-2}\nabla u(x), \nabla\zeta(x)\rangle dx$$

$$\leq 2\int_\Omega \zeta(x)|u(x+h)-u(x)|\big||\nabla u(x+h)|^{p-2}\nabla u(x+h) - |\nabla u(x)|^{p-2}\nabla u(x)\big||\nabla\zeta(x)|dx$$

To continue we need the "elementary inequalities"

$$\frac{4}{p^2}\big||b|^{\frac{p-2}{2}}b - |a|^{\frac{p-2}{2}}a\big|^2 \leq \langle |b|^{p-2}b - |a|^{p-2}a, b-a\rangle,$$

$$\big||b|^{p-2}b - |a|^{p-2}a\big| \leq (p-1)\big(|a|^{\frac{p-2}{2}} + |b|^{\frac{p-2}{2}}\big)\big||b|^{\frac{p-2}{2}}b - |a|^{\frac{p-2}{2}}a\big|$$

given in Chap. 12. We obtain

$$\frac{4}{p^2}\int_\Omega \zeta^2(x)|F(x+h)-F(x)|^2 dx$$

$$\leq 2(p-1)\int_\Omega |u(x+h)-u(x)||\nabla\zeta(x)|\big(|\nabla u(x+h)|^{\frac{p-2}{2}} + |\nabla u(x)|^{\frac{p-2}{2}}\big)\zeta(x)|F(x+h)-F(x)|dx$$

$$\leq 2(p-1)\Big\{\int_\Omega |u(x+h)-u(x)|^p|\nabla\zeta(x)|^p dx\Big\}^{\frac{1}{p}}\Big\{\int_\Omega \zeta^2(x)|F(x+h)-F(x)|^2 dx\Big\}^{\frac{1}{2}}$$

$$\cdot\Big\{\int_{\text{supp}\,\zeta} \big(|\nabla u(x+h)|^{\frac{p-2}{2}} + |\nabla u(x)|^{\frac{p-2}{2}}\big)^{\frac{2p}{p-2}} dx\Big\}^{\frac{p-2}{2p}}.$$

At the last step Hölder's inequality with the three exponents p, 2 and $2p/(p-2)$ was used; indeed, they match

$$\frac{1}{p} + \frac{1}{2} + \frac{p-2}{2p} = 1$$

as required. The last integral factor is majorized by

$$\Big(\int_\Omega |\nabla u(x+h)|^p dx\Big)^{\frac{p-2}{2p}} + \Big(\int_\Omega |\nabla u(x)|^p dx\Big)^{\frac{p-2}{2p}}$$

$$\leq 2\Big(\int_\Omega |\nabla u(x)|^p dx\Big)^{\frac{p-2}{2p}} = 2\Big(\int_\Omega |F|^2 dx\Big)^{\frac{p-2}{2p}}$$

according to Minkowski's inequality, when $|h|$ is small. Dividing out the common factor (=the square root of the integral containing $F(x+h) - F(x)$) we arrive at

$$\frac{1}{p^2}\left\{\int\limits_\Omega \zeta^2(x)\left|\frac{F(x+h)-F(x)}{h}\right|^2 dx\right\}^{\frac{1}{2}}$$

$$\leq (p-1)\left\{\int\limits_\Omega |F|^2 dx\right\}^{\frac{p-2}{2p}}\left\{\int\limits_\Omega \left|\frac{u(x+h)-u(x)}{h}\right|^p |\nabla\zeta(x)|^p dx\right\}^{\frac{1}{p}} \tag{4.4}$$

Recall the characterization of Sobolev spaces in terms of integrated difference quotients (see for example Sect. 7.11 in [GT] or [LU, Chap. 2, Lemma 4.6, p.65]. We conclude that

$$\left\{\int\limits_\Omega \left|\frac{u(x+h)-u(x)}{h}\right|^p |\nabla\zeta(x)|^p dx\right\}^{\frac{1}{p}} \leq \frac{C_n}{\text{dist}(G,\partial\Omega)}\left\{\int\limits_\Omega |\nabla u(x)|^p dx\right\}^{\frac{1}{p}}$$

Hence (4.3) yields

$$\left\{\int\limits_G \left|\frac{F(x+h)-F(x)}{h}\right|^2 dx\right\}^{\frac{1}{2}} \leq \frac{C(n,p)}{\text{dist}(G,\partial\Omega)}\left\{\int\limits_\Omega |F(x)|^2 dx\right\}^{\frac{1}{2}}$$

This is sufficient to guarantee that $F \in W^{1,2}(G)$ and the desired bound follows.   □

Remark A pretty simple proof of the Hölder continuity of u is available, when $p > n-2$ and $p \geq 2$. It is based on Theorem 4.1. The reasoning is as follows. Since the differential DF belongs to $L^2_{\text{loc}}(\Omega)$ by the theorem, Sobolev's inbedding theorem assures that $F \in L^{2n/(n-2)}_{\text{loc}}(\Omega)$, that is $\nabla u \in L^{np/(n-2)}_{\text{loc}}(\Omega)$. This summability exponent is large. Indeed

$$\frac{np}{n-2} > n \quad \text{when } p > n-2.$$

We conclude that $u \in C^\alpha_{\text{loc}}(\Omega)$, with $\alpha = 1 - (n-2)/p$, since it belongs to some $W^{1,s}_{\text{loc}}(\Omega)$ where s is greater than the dimension n.

This was the case $p \geq 2$.

In the case $1 < p < 2$ the previous proof does not work. However, an ingenious trick, mentioned in [G, Sect. 8.2], leads to a stronger result. We start with a simple fact.

Lemma 4.5 *Let $f \in L^1_{\text{loc}}(\Omega)$. Then*

$$\int\limits_\Omega \varphi(x)\frac{f(x+he_k)-f(x)}{h} dx = -\int\limits_\Omega \frac{\partial\varphi}{\partial x_k}\left(\int\limits_0^1 f(x+the_k)dt\right)dx$$

holds for all $\varphi \in C^\infty_0(\Omega)$.

Proof For a smooth function f the identity holds, because

$$\frac{\partial}{\partial x_k} \int_0^1 f(x + the_k)dt = \frac{f(x + he_k) - f(x)}{h}$$

by the infinitesimal calculus. The general case follows by approximation. □

Regarding the x_k-axis as the chosen direction, we use the abbreviation

$$\Delta^h f = \Delta^h f(x) = \frac{f(x + he_k) - f(x)}{h}$$

By the lemma the formula

$$\Delta^h(|\nabla u|^{p-2}\nabla u) = \frac{\partial}{\partial x_k} \int_0^1 |\nabla u(x + the_k)|^{p-2}\nabla u(x + the_k)dt$$

can be used in Sobolev's sense.

Theorem 4.6 *Let* $1 < p \le 2$. *If* u *is p-harmonic in* Ω, *then* $u \in W^{2,p}_{loc}(\Omega)$. *Moreover*

$$\int_D \left|\frac{\partial^2 u}{\partial x_i \partial x_j}\right|^p dx \le C_D \int_\Omega |\nabla u|^p dx$$

when $D \subset\subset \Omega$.

Proof Use formula (4.1) again. In our new notation the identity next after (4.1) can be written as

$$\int \zeta^2 \langle \Delta^h(|\nabla u|^{p-2}\nabla u), \Delta^h(\nabla u)\rangle dx$$

$$= -2 \int \zeta \Delta^h u \langle \Delta^h(|\nabla u|^{p-2}\nabla u), \nabla \zeta\rangle dx$$

$$= 2 \int \langle \int_0^1 |\nabla u(x + the_k)|^{p-2}\nabla u(x + the_k)dt, \frac{\partial}{\partial x_k}(\Delta^h u \cdot \zeta\nabla\zeta)\rangle dx$$

The last equality was based on Lemma 4.5. This was "the ingenious trick". We have

$$\frac{\partial}{\partial x_k}(\Delta^h u \cdot \zeta\nabla\zeta) = \zeta\nabla\zeta\Delta^h u_{x_k} + \Delta^h u(\zeta_{x_k}\nabla\zeta + \zeta\nabla\zeta_{x_k})$$

by direct differentiation. Let us fix a ball $B_{3R} \subset\subset \Omega$ and select a cutoff function ζ vanishing outside B_{2R}, $\zeta = 1$ in B_R, $0 \le \zeta \le 1$ such that

$$|\nabla \zeta| \le R^{-1}, \quad |D^2 \zeta| \le CR^{-2}$$

For simplicity, abbreviate

$$Y(x) = \int\limits_{B_{2R}} |\nabla u(x + the_k)|^{p-1} dt.$$

The estimate

$$\int\limits_{\Omega} \zeta^2 \langle \Delta^h(|\nabla u|^{p-2}\nabla u), \Delta^h(\nabla u)\rangle dx$$

$$\le \frac{2}{R} \int\limits_{\Omega} \zeta Y |\Delta^h u_{x_k}| dx + \frac{c}{R^2} \int\limits_{B_{2R}} |\Delta^h u| Y dx$$

(4.7)

follows. Since $1 < p < 2$, the inequality

$$\langle |b|^{p-2}b - |a|^{p-2}a, b - a\rangle \ge (p-1)|b-a|^2(1 + |a|^2 + |b|^2)^{\frac{p-2}{2}}$$

is available, see VII in Sect. 12, and we can estimate the left hand side of (4.7) from below. With the further abbreviation

$$W(x)^2 = 1 + |\nabla u(x)|^2 + |\nabla u(x + he_k)|^2$$

we write, using also $|\Delta^h(\nabla u)| \ge |\Delta^h u_{x_k}|$,

$$(p-1) \int\limits_{\Omega} \zeta^2 W^{p-2} |\Delta^h(\nabla u)|^2 dx \le \frac{2}{R} \int\limits_{\Omega} \zeta Y |\Delta^h(\nabla u)| dx$$

$$+ \frac{c}{R^2} \int\limits_{B_{2R}} |\Delta^h u| Y dx$$

The first term in the right hand member has to be absorbed (the so-called Peter-Paul Principle). To this end, let $\varepsilon > 0$ and use

$$2R^{-1}\zeta Y |\Delta^h(\nabla u)| = 2\zeta W^{(p-2)/2}|\Delta^h(\nabla u)| W^{(2-p)/2} Y R^{-1}$$

$$\le \varepsilon \zeta^2 W^{p-2} |\Delta^h(\nabla u)|^2 + \varepsilon^{-1} R^{-2} W^{2-p} Y^2.$$

For example, $\varepsilon = (p - 1)/2$ will do. The result is then

$$\frac{p-1}{2} \int_{B_R} W^{p-2} |\Delta^h(\nabla u)|^2 dx \leq \frac{2}{p-1} R^{-2} \int_{B_{2R}} W^{2-p} Y^2 dx$$

$$+ cR^{-2} \int_{B_{2R}} |\Delta^h u| Y dx.$$

Incorporating the elementary inequalities

$$|\Delta^h(\nabla u)|^p \leq W^{p-2} |\Delta^h(\nabla u)|^2 + W^p,$$
$$W^{2-p} Y^2 \leq W^p + Y^{p/(p-1)},$$
$$|\Delta^h u| Y \leq |\Delta^h u|^p + Y^{p/(p-1)},$$

the estimate takes the form

$$\int_{B_R} |\Delta^h(\nabla u)|^p dx \leq c_1 \int_{B_{2R}} W^p dx + c_2 \int_{B_{2R}} Y^{\frac{p}{p-1}} dx + c_3 \int_{B_{2R}} |\Delta^h u|^p dx$$

where the constants also depend on R. It remains to bound the three integrals as $h \to 0$. First, it is plain that

$$\int_{B_{2R}} W^p dx \leq CR^n + C \int_{B_{3R}} |\nabla u|^p dx.$$

Second, the middle integral is bounded as follows:

$$\int_{B_{2R}} Y^{\frac{p}{p-1}} dx = \int_{B_{2R}} \left(\int_0^1 |\nabla u(x + the_k)|^{p-1} dt \right)^{\frac{p}{p-1}} dx$$

$$\leq \int_{B_{2R}} \int_0^1 |\nabla u(x + the_k)|^p dt dx \leq \int_{B_{3R}} |\nabla u|^p dx$$

for h small enough. For the last integral the bound

$$\int_{B_{2R}} |\Delta^h u|^p dx \leq \int_{B_{3R}} |\nabla u|^p dx$$

follows from the characterization of Sobolev's space in terms of integrated difference quotients.

Collecting the three bounds, we have the final estimate

$$\int_{B_R} |\Delta^h(\nabla u)|^p dx \le C(n, p, R) \int_{B_{3R}} |\nabla u|^p dx$$

and the theorem follows. □

Chapter 5
On p-Superharmonic Functions

In the classical potential theory the subharmonic and superharmonic functions play a central rôle. The gravitational potential predicted by Newton's theory is the leading example. It is remarkable that the mathematical features of this linear theory are, to a great extent, preserved when the Laplacian is replaced by the p-Laplace operator or by some more general differential operator with a similar structure. Needless to say, the principle of superposition is naturally lost in this generalization. This is the modern non-linear potential theory, based on partial differential equations. This chapter is taken from [L2].

5.1 Definition and Examples

The definition is based on the *Comparison Principle*. (In passing, we mention that there is an equivalent definition used in the modern theory of viscosity solutions and the p-superharmonic functions below are precisely the *viscosity supersolutions*, cf. [JLM].)

Definition 5.1 A function $v : \Omega \to (-\infty, \infty]$ is called *p-superharmonic* in Ω, if

(i) v is lower semi-continuous in Ω
(ii) $v \not\equiv \infty$ in Ω
(iii) for each domain $D \subset\subset \Omega$ the Comparison Principle holds: if $h \in C(\overline{D})$ is p-harmonic in D and $h|\partial D \le v|\partial D$, then $h \le v$ in D

A function $u : \Omega \to [-\infty, \infty)$ is called *p-subharmonic* if $v = -u$ is p-superharmonic.

It is clear that a function is *p-harmonic* if and only if it is both p-subharmonic and p-superharmonic, but Theorem 2.16 is needed for a proof.

For $p = 2$ this is the classical definition of F. Riesz. See [R] We emphasize that not even the existence of the gradient ∇v is required in the definition. (A very attentive

© The Author(s), under exclusive license to Springer Nature Switzerland AG 2019
P. Lindqvist, *Notes on the Stationary p-Laplace Equation*,
SpringerBriefs in Mathematics, https://doi.org/10.1007/978-3-030-14501-9_5

reader might have noticed that the definition does not have a local character.) As we will learn, it exists in Sobolev's sense. For sufficiently regular p-superharmonic functions we have the following, more practical, characterization.

Theorem 5.2 *Suppose that v belongs to $C(\Omega) \cap W_{loc}^{1,p}(\Omega)$. Then the following conditions are equivalent*

(i) $\int_D |\nabla v|^p dx \leq \int_D |\nabla(v + \eta)|^p dx$ *whenever* $D \subset\subset \Omega$ *and* $\eta \in C_0^\infty(D)$ *is non-negative*

(ii) $\int \langle |\nabla v|^{p-2}\nabla v, \nabla\eta\rangle dx \geq 0$ *whenever* $\eta \in C_0^\infty(\Omega)$ *is non-negative*

(iii) v *is p-superharmonic in* Ω.

Proof The equivalence of (i) and (ii) is well-known in the Calculus of Variations. If (ii) is valid, so is (i) because

$$|\nabla(v + \eta)|^p \geq |\nabla v|^p + p\langle|\nabla v|^{p-2}\nabla v, \nabla\eta\rangle.$$

If (i) holds, then the function

$$J(\varepsilon) = \int_D |\nabla(v(x) + \varepsilon\eta(x))|^p dx$$

satisfies $J(0) \leq J(\varepsilon)$, when $\varepsilon \geq 0$. Here the domain $D \subset\subset \Omega$ contains the support of η. By the infinitesimal calculus $J'(0) \geq 0$. This is (ii).

It remains to show that (ii) and (iii) are equivalent. First, suppose that (ii) holds. Let $D \subset\subset \Omega$ and suppose that $h \in C(\overline{D})$ is p-harmonic in D and $h \leq v$ on ∂D. The test function[1]

$$\eta = \max\{h - v, 0\}$$

produces the inequality

$$\int_{v\leq h} |\nabla v|^p dx \leq \int_{v\leq h} \langle|\nabla v|^{p-2}\nabla v, \nabla h\rangle dx$$

$$\leq \left\{\int_{v\leq h} |\nabla v|^p dx\right\}^{1-\frac{1}{p}} \left\{\int_{v\leq h} |\nabla h|^p dx\right\}^{\frac{1}{p}}.$$

Hence

$$\int_{v<h} |\nabla v|^p dx \leq \int_{v<h} |\nabla h|^p dx.$$

---

[1]If $\int_D |\nabla h|^p\, dx = \infty$, use the test function $\eta(x) = \max\{h(x) - \varepsilon - v(x), 0\}$ so that $\int_D |\nabla\eta|^p\, dx$ converges.

In other words, v is a minimizer in (each component of) the open set $\{v < h\}$. The boundary values are $v = h$. The minimizer is unique and so $v = h$ in this set. This contradiction proves that $v \geq h$. Thus (ii) or (i) implies (iii).

The proof of the sufficiency of (iii) seems to require the introduction of an obstacle problem. It will be given in Corollary 5.9, which does not rely on "(iii) $\Rightarrow$ (ii)" when it comes to its proof. □

Remark The continuity of v is not needed for the equivalency of (i) and (ii). The whole theorem holds for lower semicontinuous functions in the Sobolev space. More could be said about this.

It is instructive to consider some examples. The one-dimensional situation is enlighting. The p-harmonic functions in one variable are just the line segments $h(x) = ax + b$. Now p has no bearing. The p-superharmonic functions are exactly the concave functions of one variable. The comparison principle is the familiar "arc above chord" condition. In several dimensions, the concave functions are p-superharmonic, simultaneously for all p, but there are many more of them.

The leading example of a p-superharmonic function is

$$(n - p)|x|^{\frac{p-n}{p-1}} \quad (p \neq n), \quad -\log|x| \quad (p = n),$$

usually multiplied by a positive normalizing constant. Outside the origin the function is p-harmonic. Notice that the function is not of class $W^{1,p}(\Omega)$, if Ω contains the origin. Therefore it is not a weak supersolution in the sense of Definition 2.12! See [KV] about isolated singularities in general. We cannot resist mentioning that, although the principle of superposition is not valid, the function

$$v(x) = \int_\Omega \frac{\varrho(y)dy}{|x - y|^{(n-p)/(p-1)}} \quad (1 < p < n) \tag{5.3}$$

is, indeed, p-superharmonic for $\varrho(y) \geq 0$. This follows from an interesting calculation by Crandall and Zhang done for the corresponding Riemann sums, cf [CZ]. See also [Br] and [LiM3]. Of course, this remarkable representation formula cannot directly give all the p-superharmonic functions.

Theorem 5.4 (Hadamard's Three-Spheres Theorem) *If u is p-subharmonic in the ring domain $r < |x| < R$ and continuous up to the boundary spheres, then*

$$u(x) \leq \frac{|x|^\alpha - R^\alpha}{r^\alpha - R^\alpha} M(r) + \frac{r^\alpha - |x|^\alpha}{r^\alpha - R^\alpha} M(R), \quad M(r) = \max_{\partial B(0,r)} \{u\},$$

where $\alpha = (p - n)/(p - 1)$, $p \neq n$. (When $p = n$, the powers become logarithms.)

Proof The right hand-side is a p-harmonic function. The inequality is valid on the boundary spheres. Thus the result follows by comparison. □

It is useful that the pointwise minimium of two p-superharmonic functions is again p-superharmonic as a direct consequence of the definition.

Before going further we had better make a simple comment. Assumption (ii) in Definition 5.1 means that v is finite at least at one point. In fact, it follows easily that the set $\{v < \infty\}$ is dense in Ω. (As we will later see, $v < \infty$ a.e..)

Proposition 5.5 *If v is p-superharmonic in Ω, then the set where $v = \infty$ does not contain any ball.*

Proof Suppose to begin with that $v \geq 0$ in Ω. Assume that $v \equiv +\infty$ in some ball $\overline{B}_r = B(x_0, r)$ and that $B_R = B(x_0, R) \subset\subset \Omega$, where $R > r$. We claim that $v \equiv +\infty$ also in the larger ball B_R. The function

$$h(x) = \frac{\int\limits_{|x-x_0|}^{R} t^{-(n-1)/(p-1)} dt}{\int\limits_{r}^{R} t^{-(n-1)/(p-1)} dt}$$

is p-harmonic when $x \neq x_0$, in particular it is p-harmonic in the annulus $r < |x - x_0| < R$. It takes the boundary values 0 on ∂B_R and 1 on ∂B_r. Consider the p-harmonic function $kh(x)$. The comparison principle shows that

$$v(x) \geq kh(x), \quad k = 1, 2, 3, \ldots$$

in the annulus. We conclude that $v \equiv \infty$ in the annulus. In other words $v \equiv \infty$ in B_R.

To get rid of the restriction $v \geq 0$, we consider the function $v - \inf v$ instead of v. Again the conclusion is that $v|B_R \equiv \infty$ if $v|B_r \equiv \infty$.

Repeating the procedure through a suitable chain of balls, we finally arrive at the contradiction $v \equiv \infty$ in Ω.   □

5.2   The Obstacle Problem and Approximation

As we have seen, the p-harmonic functions come from a minimization problem in the Calculus of Variations. If one adds a restriction on the admissible functions, when minimizing, weak supersolutions of the p-harmonic equation are produced. The restrictive condition is nothing more than that the functions have to lie above a given function, which acts as a fixed *obstacle*.

Suppose, as usual, that Ω is a bounded domain in $\mathbf{R}^n$. Given a function $\psi \in C(\Omega) \cap W^{1,p}(\Omega)$ we consider the problem of minimizing the integral

$$\int\limits_{\Omega} |\nabla v|^p dx$$

among all functions in the class

$$\mathcal{F}_\psi(\Omega) = \{v \in C(\Omega) \cap W^{1,p}(\Omega) | v \geq \psi \text{ in } \Omega \text{ and } v - \psi \in W_0^{1,p}(\Omega)\}.$$

This is the obstacle problem with ψ acting as an obstacle from below. Also the boundary values are prescribed by ψ. (One could also allow other boundary values, but we do not discuss this variant.)

Theorem 5.6 *Given $\psi \in C(\Omega) \cap W^{1,p}(\Omega)$, there exists a unique minimizer v_ψ in the class $\mathcal{F}(\Omega)$, i.e.*

$$\int_\Omega |\nabla v_\psi|^p dx \leq \int_\Omega |\nabla v|^p dx$$

for all similar v. The function v_ψ is p-superharmonic in Ω and p-harmonic in the open set $\{v_\psi > \psi\}$. If in addition, Ω is regular enough and $\psi \in C(\overline{\Omega})$, then also $v_\psi \in C(\overline{\Omega})$ and $v_\psi = \psi$ on $\partial\Omega$.

Proof The existence of a unique minimizer is easily established, except for the continuity; only some functional analysis is needed. Compare with the problem without obstacle in section 2. It is the continuity that is difficult to prove in the case $1 < p \leq n$. We refer to [MZ] for the proof of the continuity of v_ψ.

Next we conclude that

$$\int_\Omega \langle |\nabla v_\psi|^{p-2} \nabla v_\psi, \nabla\eta\rangle dx \geq 0 \tag{5.7}$$

when $\eta \in C_0^\infty(\Omega)$, $\eta \geq 0$, according to Theorem 5.2, which also assures that v_ψ is p-superharmonic in Ω.

We have come to the important property that v_ψ is p-harmonic in the set where the obstacle does not hinder, say

$$S = \{x \in \Omega | v_\psi(x) > \psi(x)\}.$$

In fact, we can conclude that (5.7) is valid for all $\eta \in C_0^\infty(\Omega)$, positive or not, satisfying

$$v_\psi(x) + \varepsilon\eta(x) \geq \psi(x)$$

for small $\varepsilon > 0$. Consequently, we can remove the sign restriction on η in the set S. Indeed, if $\eta \in C_0^\infty(S)$ it suffices to consider ε so small that

$$\varepsilon\|\eta\|_\infty \leq \min(v_\psi - \psi)$$

the minimum being taken over the support of η. Here η can take also negative values. We conclude that v is p-harmonic in S.

For the question about classical boundary values in regular domains we refer to [F]. □

I take myself the liberty to hint that it is a good exercise to work out the previous proof in the one-dimensional case, where no extra difficulties obscure the matter, and pictures can be drawn.

Remark More advanced regularity theorems hold for the solution. If the obstacle is smooth, then v_ψ is of class $C_{loc}^{1,\alpha}(\Omega)$. Of course, the regularity cannot be any better than for general p-harmonic functions. We refer to [CL, S, L3] about the gradient ∇v_ψ.

In the sequel we will use a sequence of obstacles to study the differentiability properties of p-superharmonic functions. The proof that an arbitrary p-superharmonic function v has Sobolev derivatives requires several steps:

(1) v is pointwise approximated from below by smooth functions ψ_j.
(2) The obstacle problem with ψ_j acting as an obstacle is solved. It turns out that

$$\psi_j(x) \le v_{\psi_j}(x) \le v(x).$$

(3) Since the v_{ψ_j}'s are supersolutions, they satisfy expedient a priori estimates.
(4) The a priori estimates are passed over to $v = \lim v_{\psi_j}$, first in the case when v is bounded.
(5) For an unbounded v one goes via the bounded p-superharmonic functions $\min\{v(x), k\}$ and an estimate free of k is reached at the end.

To this end, we assume that v is p-superharmonic in Ω. Because of the lower semicontinuity of v, there exists an increasing sequence of functions $\psi_j \in C^\infty(\Omega)$ such that

$$\psi_1(x) \le \psi_2(x) \le \cdots \le v(x), \quad \lim_{j\to\infty} \psi_j(x) = v(x)$$

at each $x \in \Omega$. Next, fix a regular bounded domain $D \subset\subset \Omega$. Let $v_j = v_{\psi_j}$ denote the solution of the obstacle problem in D, the function ψ_j acting as an obstacle. Thus $v_j \in \mathcal{F}_{\psi_j}(D)$ and $v_j \ge \psi_j$ in D. We claim that

$$v_1 \le v_2 \le \dots, \quad \psi_j \le v_j \le v$$

pointwise in D. To see that $v_j \le v$, we notice that this is true except possibly in the open set $A_j = \{v_j > \psi_j\}$, where the obstacle does not hinder. By Theorem 5.6 v_j is p-harmonic in A_j (provided that A_j is not empty) and on the boundary ∂A_j we know that $v_j = \psi_j$. Hence $v_j \le v$ on ∂A_j and so the comparison principle, which v is assumed to obey, implies that $v_j \le v$ in A_j. This was the main point in the proof, here the comparison principle was used. We have proved that $v_j \le v$ at each point in D.

The inequalities $v_j \le v_{j+1}$, $j = 1, 2, 3, \dots$, have a similar proof, because v_{j+1} satisfies the comparison principle according to Theorem 5.6.

We have established the first part of the next theorem.

Theorem 5.8 *Suppose that v is a p-superharmonic function in the domain Ω. Given a subdomain $D \subset\subset \Omega$ there are such p-superharmonic functions $v_j \in C(\overline{D}) \cap W^{1,p}(D)$ that*

$$v_1 \leq v_2 \leq \ldots \text{ and } v = \lim_{j \to \infty} v_j$$

at each point in D. If, in addition, v is (locally) bounded from above in Ω, then also $v \in W^{1,p}_{loc}(\Omega)$, and the approximants v_j can be chosen so that

$$\lim_{j \to \infty} \int_D |\nabla(v - v_j)|^p dx = 0.$$

Proof Fix D and choose a regular domain D_1, $D \subset\subset D_1 \subset\subset \Omega$. By the previous construction there are p- superharmonic functions v_j in D_1 such that $v_1 \leq v_2 \leq \ldots$, $v_j \to v$ pointwise in D_1 and $v_j \in C(D_1) \cap W^{1,p}(D_1)$.

For the second part of the theorem we know that

$$C = \sup_{D_1} v - \inf_{D_1} \psi_1 < \infty$$

if v is locally bounded. Theorem 5.2 and a simple modification of Lemma 2.9 to include weak supersolutions lead to the bound

$$\int_D |\nabla v_j|^p dx \leq p^p C^p \int_{D_1} |\nabla \zeta|^p dx = M \quad (j = 1, 2, 3, \ldots).$$

By a standard compactness argument $v \in W^{1,p}(D)$ and $\|\nabla v\|_{L^p(D)} \leq M$. For a subsequence we have that $\nabla v_k \rightharpoonup \nabla v$ weakly in $L^p(D)$. We also conclude that $v \in W^{1,p}_{loc}(\Omega)$.

To establish the strong convergence of the gradients, it is enough to show that

$$\lim_{j \to \infty} \int_{B_r} |\nabla v - \nabla v_j|^p dx = 0$$

whenever B_r is such a ball in D that the concentric ball B_{2r} (with double radius) is comprised in D_1. As usual, let $\zeta \in C_0^\infty(B_{2r})$, $0 \leq \zeta \leq 1$ and $\zeta = 1$ in B_r. Next, use the non-negative test function $\eta_j = \zeta(v - v_j)$ in the equation

$$\int_{B_{2r}} \langle |\nabla v_j|^{p-2} \nabla v_j, \nabla \eta_j \rangle dx \geq 0$$

to find that

$$J_j = \int_{B_{2r}} \langle |\nabla v|^{p-2}\nabla v - |\nabla v_j|^{p-2}\nabla v_j, \nabla(\zeta(v-v_j)))dx$$

$$\leq \int_{B_{2r}} \langle |\nabla v|^{p-2}\nabla v, \nabla(\zeta(v-v_j)))dx$$

By the weak convergence of the gradients

$$\limsup_{j\to\infty} J_j \leq 0.$$

We split J_j in two parts:

$$J_j = \int_{B_{2r}} \zeta\langle |\nabla v|^{p-2}\nabla v - |\nabla v_j|^{p-2}\nabla v_j, \nabla v - \nabla v_j\rangle dx$$

$$+ \int_{B_{2r}} (v-v_j)\langle |\nabla v|^{p-2}\nabla v - |\nabla v_j|^{p-2}\nabla v_j, \nabla\zeta\rangle dx$$

The last integral is bounded in absolute value by the majorant

$$\left\{ \int_{B_{2r}} (v-v_j)^p dx \right\}^{\frac{1}{p}} \left\{ \left(\int_{B_{2r}} |\nabla v|^p dx \right)^{1-\frac{1}{p}} + \left(\int_{B_{2r}} |\nabla v_j|^p dx \right)^{1-\frac{1}{p}} \right\} \max |\nabla\zeta|$$

$$\leq 2M^{1-\frac{1}{p}} \max |\nabla\zeta| \left\{ \int_{B_{2r}} (v-v_j)^p dx \right\}^{\frac{1}{p}}$$

and hence it approaches zero as $j \to \infty$. Collecting results, we see that

$$\lim_{j\to\infty} \int_{B_{2r}} \zeta\langle |\nabla v|^{p-2}\nabla v - |\nabla v_j|^{p-2}\nabla v_j, \nabla v - \nabla v_j\rangle dx \leq 0$$

at least for a subsequence. The integrand is non-negative. For $p \geq 2$ we can use the inequality

$$\zeta\langle |\nabla v|^{p-2}\nabla v - |\nabla v_j|^{p-2}\nabla v_j, \nabla v - \nabla v_j\rangle \geq 2^{2-p}|\nabla v - \nabla v_j|^p$$

in B_r to conclude the proof. The reader might find it interesting to complete the proof for $1 < p < 2$. □

With this approximation theorem it is easy to prove that bounded p-superharmonic functions are weak supersolutions. Also the opposite statement is true, provided that the issue about semicontinuity be properly handled.

Corollary 5.9 *Suppose that v is p-superharmonic and locally bounded in Ω. Then $v \in W^{1,p}_{loc}(\Omega)$ and v is a weak supersolution:*

$$\int_{\Omega} \langle |\nabla v|^{p-2}\nabla v, \nabla \eta \rangle dx \geq 0$$

for all non-negative $\eta \in C_0^{\infty}(\Omega)$.

Proof We have to justify the limit procedure

$$\int_{\Omega} \langle |\nabla v|^{p-2}\nabla v, \nabla \eta \rangle dx = \lim_{j \to \infty} \int_{\Omega} \langle |\nabla v_j|^{p-2}\nabla v_j, \nabla \eta \rangle dx \geq 0$$

where the v_j's are the approximants in Theorem 5.8. By their construction they solve an obstacle problem and hence they are weak supersolutions (Theorem 5.2). In the case $p \geq 2$, one can use the inequality

$$\left| |\nabla v|^{p-2}\nabla v - |\nabla v_j|^{p-2}\nabla v_j \right| \leq$$
$$(p-1)|\nabla v - \nabla v_j|(|\nabla v|^{p-2} + |\nabla v_j|^{p-2})$$

and then apply Hölder's inequality. In the case $1 < p \leq 2$ one has directly that

$$\left| |\nabla v|^{p-2}\nabla v - |\nabla v_j|^{p-2}\nabla v_j \right| \leq \gamma(p)|\nabla v - \nabla v_j|^{p-1}.$$

The strong convergence in Theorem 5.8 is needed in both cases. $\square$

We make a discursion and consider the convergence of an increasing sequence of p-harmonic functions.

Theorem 5.10 (Harnack's convergence theorem) *Suppose that h_j is p-harmonic and that*

$$0 \leq h_1 \leq h_2 \leq \ldots, \quad h = \lim h_j$$

pointwise in Ω. Then, either $h \equiv \infty$ or h is a p-harmonic function in Ω.

Proof Recall the Harnack inequality (Theorem 2.20)

$$h_j(x) \leq Ch_j(x_0), \quad j = 1, 2, 3, \ldots$$

valid for each $x \in B(x_0, r)$, when $B(x_0, 2r) \subset\subset \Omega$. The constant C is independent of the index j. If $h(x_0) < \infty$ at some point x_0, then $h(x) < \infty$ at each $x \in \Omega$. This we can deduce using a suitable chain of balls. It also follows that h is locally bounded in this case.

The Cacciopppoli estimate

$$\int_{B_r} |\nabla h_j|^p dx \leq Cr^{-p} \int_{B_{2r}} |h_j|^p dx \leq Cr^{-p} \int_{B_{2r}} |h|^p dx$$

$$\leq c_1 C^p r^{n-p} h(x_0)^p$$

allows us to conclude that $h \in W^{1,p}_{loc}(\Omega)$. Finally,

$$\int_{\Omega} \langle |\nabla h|^{p-2} \nabla h, \nabla \eta \rangle dx = \lim_{j \to \infty} \int_{\Omega} \langle |\nabla h_j|^{p-2} \nabla h_j, \nabla \eta \rangle dx = 0$$

for each $\eta \in C_0^\infty(\Omega)$ follows from a repetition of the corresponding argument in the proof of Theorem 5.8.   □

5.3   Infimal Convolutions

Instead of using the approximation with solutions of obstacle problems, as in the previous subsection, one can directly use so-called *infimal convolutions*. They inherit the comparison principle. See [LiM1].

Suppose that v is lower semicontinuous and bounded in Ω :

$$0 \leq v(x) \leq L$$

and define

$$v_\varepsilon(x) = \inf_{y \in \Omega} \left\{ v(y) + \frac{|x-y|^2}{2\varepsilon} \right\}, \quad \varepsilon > 0. \tag{5.11}$$

Then

- $v_\varepsilon(x) \nearrow v(x)$ as $\varepsilon \to 0+$
- $v_\varepsilon(x) - |x|^2/2\varepsilon$ is locally concave in $\mathbf{R}^n$
- v_ε is locally Lipschitz continuous in Ω
- The Sobolev gradient ∇v_ε exists and belongs to $L^\infty_{loc}(\Omega)$
- The second derivatives $\mathbb{D}^2 v_\varepsilon$ exist in Alexandrov's sense. See Sect. 9

The assertion about the Sobolev derivatives follow from Rademacher's theorem about Lipschitz functions, see [EG]. A most remarkable feature is that the infimal convolutions preserve the property of p-superharmonicity.

Lemma 5.12 *If v is a p-superharmonic function in Ω and $0 \leq v \leq L$, the infimal convolution v_ε is a p-superharmonic function in the open subset of Ω where*

$$\mathrm{dist}(x, \partial\Omega) > \sqrt{2L\varepsilon}.$$

Similarly, the local weak supersolutions of the p-Laplace equation are preserved.

Proof We assume that v is p-superharmonic in

$$\Omega_\varepsilon = \left\{ x \in \Omega | \operatorname{dist}(x, \partial\Omega) > \sqrt{2L\varepsilon} \right\}.$$

First, notice that for x in Ω_ε, the infimum is attained at some point $y = x^*$ comprised in Ω. The possibility that x^* escapes to the boundary of Ω is hindered by the inequalities

$$\frac{|x - x^*|^2}{2\varepsilon} \leq \frac{|x - x^*|^2}{2\varepsilon} + v(x^*) = v_\varepsilon(x) \leq v(x) \leq L,$$
$$|x - x^*| \leq \sqrt{2L\varepsilon} < \operatorname{dist}(x, \partial\Omega).$$

This explains why the domain shrinks a little.

We have to verify the comparison principle for v_ε in an arbitrary subdomain $D \subset\subset \Omega_\varepsilon$. Suppose that $h \in C(\overline{D})$ is a p-harmonic function such that $v_\varepsilon(x) \geq h(x)$ on the boundary ∂D or, in other words,

$$\frac{|x - y|^2}{2\varepsilon} + v(y) \geq h(x) \quad \text{when} \quad x \in \partial D, \ y \in \Omega.$$

Thus, writing $y = x + z$, we have

$$w(x) \equiv v(x + z) + \frac{|z|^2}{2\varepsilon} \geq h(x), \quad x \in \partial D$$

whenever z is a small fixed vector. But also $w = w(x)$ is a p-superharmonic function in Ω_ε. By the comparison principle, $w(x) \geq h(x)$ in the whole D. Given any point x_0 in D, we may choose $z = x_0^* - x_0$. This yields $v_\varepsilon(x_0) \geq h(x_0)$. Since x_0 was arbitrary, we have verified that

$$v_\varepsilon(x) \geq h(x) \quad \text{when} \quad x \in D.$$

This concludes the proof of the comparison principle.

The statement about (weak) p-supersolutions is immediate. □

5.4 The Poisson Modification

This subsection, based on [GLM], is devoted to a simple but useful auxiliary tool, generalizing Poisson's formula in the linear case $p = 2$. The so-called Poisson modification of a p-superharmonic function v is needed for instance in connexion with Perron's method. Given a regular subdomain $D \subset\subset \Omega$ it is defined as the function

$$V = P(v, D) = \begin{cases} v \text{ in } \Omega \backslash D \\ h \text{ in } D \end{cases}$$

where h is the p-harmonic function in D with boundary values v on ∂D. One verifies easily that $V \leq v$ and that V is p-superharmonic, if the original v is continuous. Otherwise, the interpretation of $h = v$ on ∂D requires some extra considerations. In the event that v is merely semicontinuous one goes via the approximants v_j in Theorem 5.8 and defines

$$V = \lim V_j = \lim P(v_j, D)$$

where we have tacitly assumed that $v_j \rightarrow v$ in the whole Ω (here this is no restriction). Now we use the Harnack convergence theorem (Theorem 5.10) on the functions h_j to conclude that the limit function $h = \lim h_j$ is p-harmonic in D. (Since $h_j \leq v_j \leq v$ the case $h \equiv \infty$ is out of the question. Also the situation $h_j \geq 0$ is easy to arrange by adding a constant to v.) With this h it is possible to verify that V is p-superharmonic. It is the limit of an increasing sequence of p-superharmonic functions.

Proposition 5.13 *Suppose that v is p-superharmonic in Ω and that $D \subset\subset \Omega$. Then the Poisson modification $V = P(v, D)$ is p-superharmonic in Ω, p-harmonic in D, and $V \leq v$. Moreover, if v is locally bounded, then*

$$\int_G |\nabla V|^p dx \leq \int_G |\nabla v|^p dx$$

for $D \subset G \subset\subset \Omega$.

Proof It remains to prove the minimization property. This follows from the obvious property

$$\int_G |\nabla V_j|^p dx \leq \int_G |\nabla v_j|^p dx.$$

In fact, the case $G = D$ is the relevant one. $\square$

5.5   Summability of Unbounded p-Superharmonic Functions

We have seen that the so-called polar set

$$\Xi = \{x \in \Omega | v(x) = \infty\}$$

of a p-superharmonic function v cannot contain any open set (Proposition 5.5). Much more can be assured. Ξ is empty, when $p > n$, and it has Lebesgue measure zero is all cases. The key is to study the p-superharmonic functions

$$v_k = v_k(x) = \min\{v(x), k\}, \quad k = 1, 2, 3, \dots.$$

Since they are locally bounded, they satisfy the inequality

$$\int_\Omega \langle |\nabla v_k|^{p-2} \nabla v_k, \nabla \eta \rangle dx \geq 0$$

for each non-negative $\eta \in C_0^\infty(\Omega)$ and so the estimates for weak supersolutions are available.

Theorem 5.14 *If v is p-superharmonic in Ω, then*

$$\int_D |v|^q dx < \infty$$

whenever $D \subset\subset \Omega$ and $0 \leq q < n(p-1)/(n-p)$ in the case $1 < p \leq n$. In the case $p > n$ the function v is continuous.

Proof Because the theorem is of a local nature, we may assume that $v > 0$ by adding a constant. Then also $v_k > 0$.

First, let $1 < p < n$. According to Corollary 3.18

$$\left\{ \fint_{B_r} v_k^q dx \right\}^{\frac{1}{q}} \leq C(p, n, q) \operatorname*{ess\,inf}_{B_r} v_k$$

whenever $q < n(p-1)/(n-p)$ and $B_{2r} \subset\subset \Omega$. The constant is independent of the index k. Since $v_k \leq v$ we obtain

$$\left(\fint_{B_r} v^q dx \right)^{\frac{1}{q}} \leq C(p, n, q) \operatorname*{ess\,inf}_{B_r} v$$

It remains to prove that

$$\operatorname*{ess\,inf}_{B_r} v < \infty$$

This is postponed till Theorem 5.15, the proof of which does not rely upon the present section.

Next, consider the case $p > n$. Here the situation $v(x) = \infty$ for a.e. x will be excluded without evoking Theorem 5.15. The estimate

$$\int_\Omega \zeta^p |\nabla \log v_k|^p dx \leq \left(\frac{p}{p-1} \right)^p \int_\Omega |\nabla \zeta|^p dx$$

in Lemma 2.14 yields, as usual,

$$\|\nabla \log v_k\|_{L^p(B_r)} \le C_1 r^{(n-p)/p}$$

if $B_{2r} \subset \Omega$. According to (2.14)

$$|\log v_k(x) - \log v_k(y)| \le C_2 |x - y|^{(p-n)/p} \|\nabla \log v_k\|_{L^p(B_r)}$$

when $x, y \in B_r$. It follows that $v_k(x) \le K v_k(y)$ and

$$v(x) \le K v(y)$$

when $x, y \in B_r$ and $B_{2r} \subset \Omega$; $K = e^{C_1 C_2}$. Thus we have proved the Harnack inequality for v.

We can immediately conclude that $v(x) < \infty$ at each point in Ω, because there is at least one such point. As we know, the Harnack inequality implies continuity. In fact, $v \in C_{\text{loc}}^\alpha(\Omega)$.

Finally, we have the borderline case $p = n$. It requires some special considerations. We omit the proof that $v \in L_{\text{loc}}^q(\Omega)$ for each $q < \infty$. □

Remark The previous theorem has been given a remarkable proof by T. Kilpeläinen and J. Malý, cf [KM1]. The use of an ingenious test function makes it possible to avoid the Moser iteration.

5.6   About Pointwise Behaviour

Although we know that $v < \infty$ in a dense subset, the conclusion that ess inf $v < \infty$ requires some additional considerations. We will prove a result about pointwise behaviour from which this follows. In order to appreciate the following investigation we should be aware of that in the linear case $p = 2$ there exists a superharmonic function v defined in $\mathbf{R}^n$ such that $v(x) = +\infty$ when all the coordinates of x are rational numbers, yet $v < \infty$ a.e.. (Actually, the polar set contains more points, since it has to be a G_δ-set.) The example is

$$v(x) = \sum_q \frac{c_q}{|x - q|^{n-2}},$$

where the $c_q > 0$ are chosen to create convergence. It is astonishing that this function has Sobolev derivatives! A similar "monster" can be constructed for $1 < p < n$.

Recall that a p-superharmonic function v is lower semicontinuous. Thus

$$v(x) \le \liminf_{y \to x} v(y) \le \operatorname{ess} \liminf_{y \to x} v(y)$$

at each point $x \in \Omega$. "Essential limes inferior" means that any set of n-dimensional Lebesgue measure zero can be neglected, when limes inferior is calculated. The definition is given in [Brelot, II.5]. In fact, the reverse inequality also holds.

Theorem 5.15 *If v is p-superharmonic in Ω, then*

$$v(x) = \operatorname{ess} \lim_{y \to x} \inf v(y)$$

at each point x in Ω.

The following lemma is the main step in the proof. A pedantic formulation cannot be avoided.

Lemma 5.16 *Suppose that v is p-superharmonic in Ω. If $v(x) \leq \lambda$ at each point x in Ω and if $v(x) = \lambda$ for a.e. x in Ω, then $v(x) = \lambda$ at each x in Ω.*

Proof The idea is that v is its own Poisson modification and for a continuous function the theorem is obvious. Therefore fix a regular subdomain $D \subset\subset \Omega$ and consider the Poisson modification $V = P(v, D)$. We have

$$V \leq v \leq \lambda$$

everywhere. We claim that $V = \lambda$ at each point in D. Since v is locally bounded it is a weak supersolution and as such it belongs to $W^{1,p}_{loc}(\Omega)$. According to Proposition 5.13

$$\int_G |\nabla V|^p dx \leq \int_G |\nabla v|^p dx = \int_G |\nabla \lambda|^p dx = 0$$

for $D \subset G \subset\subset \Omega$. Hence $\nabla V = 0$ and so V is constant in G. It follows that $V = \lambda$ a.e. in G. But in D the function V is p-harmonic. It follows that $V(x) = \lambda$ at each point x in D. Since D was arbitrary, the theorem follows. $\square$

Lemma 5.17 *If v is p-superharmonic in Ω and if $v(x) > \lambda$ for a.e. x in Ω, then $v(x) \geq \lambda$ for every x in Ω.*

Proof If $\lambda = -\infty$ there is nothing to prove. Applying Lemma 5.16 to the p-superharmonic function defined by

$$\min\{v(x), \lambda\}$$

we obtain the result in the case $\lambda > -\infty$. $\square$

Proof of Theorem 5.15: Fix any x in Ω. We must show that

$$\lambda = \operatorname{ess} \lim_{y \to x} \inf v(y) \leq v(x).$$

Given any $\varepsilon > 0$, there is a radius $\delta > 0$ such that $v(y) > \lambda - \varepsilon$ for a.e. $y \in B(x, \delta)$, where δ is small enough. By Lemma 5.17 $v(y) \geq \lambda - \varepsilon$ for each $y \in B(x, \delta)$. In particular, $v(x) \geq \lambda - \varepsilon$. Because $\varepsilon > 0$ was arbitrary, we have established that $\lambda \leq v(x)$.

5.7 Summability of the Gradient

We have seen that locally *bounded* p-superharmonic functions are of class $W^{1,p}_{loc}(\Omega)$. They have first order Sobolev derivatives. For unbounded functions the summability exponent p has to be decreased, but it is important that the exponent can be taken $\geq p - 1$. The following fascinating theorem is easy to prove at this stage.

Theorem 5.18 *Suppose that v is a p-superharmonic function defined in the domain Ω in $\mathbf{R}^n$, $p > 2 - \frac{1}{n}$. Then the Sobolev derivative*

$$\nabla v = \left(\frac{\partial v}{\partial x_1}, \dots, \frac{\partial v}{\partial x_n} \right)$$

exists and the local summability result

$$\int_D |\nabla v|^q dx < \infty, \quad D \subset\subset \Omega,$$

holds whenever $0 < q < n(p-1)/(n-1)$ in the case $1 < p \leq n$ and $q = p$ in the case $p > n$. Furthermore,

$$\int_\Omega \langle |\nabla v|^{p-2} \nabla v, \nabla \phi \rangle \, dx \geq 0$$

whenever $\phi \geq 0$ $\phi \in C_0^\infty(\Omega)$.

Remark The fundamental solution

$$|x|^{(p-n)/(p-1)} \quad (p < n), \quad -\log|x| \quad (p = n)$$

shows that the exponent q is sharp. The case $p = 2$ can be read off from the Riesz representation formula. The restriction $p > 2 - \frac{1}{n}$ is not essential but guarantees that one can take $q \geq 1$. The interpretation of ∇v would demand some care if $1 < p \leq 2 - \frac{1}{n}$.

Proof Suppose first that $v \geq 1$. Fix $D \subset\subset \Omega$ and $q < n(p-1)/(n-1)$. The cutoff functions

$$v_k = \min\{v, k\}, \quad k = 1, 2, 3, \dots,$$

are bounded p-superharmonic functions and by Corollary 5.9 they are weak super-solutions. Use the test function $\eta = \zeta^p v_k^{-\alpha}$, $\alpha > 0$, in the equation

$$\int_\Omega \langle |\nabla v_k|^{p-2} \nabla v_k, \nabla \eta \rangle dx \geq 0$$

to obtain

$$\int_\Omega \zeta^p v_k^{-1-\alpha} |\nabla v_k|^p dx \leq \left(\frac{p}{\alpha}\right)^p \int_\Omega v_k^{p-1-\alpha} |\nabla \zeta|^p dx.$$

Here $\zeta \in C_0^\infty(\Omega)$, $0 \leq \zeta \leq 1$ and $\zeta = 1$ in D. By Hölder's inequality

$$\int_D |\nabla v_k|^q dx = \int_D v_k^{(1+\alpha)q/p} |v_k^{-(1+\alpha)/p} \nabla v_k|^q dx$$

$$\leq \left\{ \int_D v_k^{(1+\alpha)q/(p-q)} dx \right\}^{1-\frac{q}{p}} \left\{ \int_D v_k^{-1-\alpha} |\nabla v_k|^p dx \right\}^{\frac{q}{p}}$$

$$\leq \left(\frac{p}{\alpha}\right)^q \left\{ \int_D v^{(1+\alpha)q/(p-q)} dx \right\}^{1-\frac{q}{p}} \left\{ \int_\Omega v^{p-1-\alpha} |\nabla \zeta|^p dx \right\}^{\frac{q}{p}}$$

for any small $\alpha > 0$. A calculation shows that

$$\frac{q}{p-q} < \frac{n(p-1)}{n-p}$$

and hence we can fix α so that also

$$\frac{(1+\alpha)q}{p-q} < \frac{n(p-1)}{n-p}.$$

Inspecting the exponents we find out that, in virtue of Theorem 5.14, the sequence $\|\nabla v_k\|_{L^q(D)}$, $k = 1, 2, 3, \ldots$ is bounded. A standard compactness argument shows that ∇v exists in D and

$$\int_D |\nabla v|^q dx \leq \lim_{k \to \infty} \int_D |\nabla v_k|^q dx.$$

Since D was arbitrary, we conclude that $v \in W_{\text{loc}}^{1,q}(\Omega)$.

Finally, the restriction $v \geq 1$ is locally removed by adding a constant to v. This concludes our proof. $\qquad\square$

The equation with measure data. The equation

$$\nabla\cdot(|\nabla v|^{p-2}\nabla v) = -C_{n,p}\delta$$

is satisfied by the fundamental solution

$$|x|^{(n-p)/(p-1)} \quad (p \neq n) \qquad \log(|x|), \quad (p = n)$$

in the distributional sense:

$$\int_{\mathbf{R}^n} \langle |\nabla v|^{p-2}\nabla v, \nabla\phi\rangle \, dx = \phi(0)$$

whenever $\phi \in C_0^\infty(\mathbf{R}^n)$. Thus Dirac's delta is the right-hand side for the fundamental solution. Theorem 5.18 enables us to produce a Radon measure for each p-superharmonic function:

$$\nabla\cdot(|\nabla v|^{p-2}\nabla v) = -\mu.$$

Recall that the summability exponent $q > p - 1$ for the gradient. This is decisive.

Theorem 5.19 *Let v be a p-superharmonic function in Ω. Then there exists a nonnegative Radon measure μ such that*

$$\int_\Omega \langle |\nabla v|^{p-2}\nabla v, \nabla\phi\rangle \, dx = \int_\Omega \phi \, d\mu$$

for all $\phi \in C_0^\infty(\mathbf{R}^n)$.

Proof We already know that v and ∇v belong to $L_{\text{loc}}^{p-1}(\Omega)$. In order to use Riesz's Representation Theorem we define the linear functional

$$\Lambda_v : C_0^\infty(\Omega) \to \mathbf{R},$$

$$\Lambda_v(\phi) = \int_\Omega \langle |\nabla v|^{p-2}\nabla v, \nabla\phi\rangle \, dx.$$

Now $\Lambda_v(\phi) \geq 0$ for $\phi \geq 0$ according to Theorem 5.18. Thus the functional is positive and the existence of the Radon measure follows from Riesz's theorem, cf. [EG, 1.8]. $\qquad\qquad\Box$

Some further results can be found in [KM1]. See also [KKT] and [KM].

Chapter 6
Perron's Method

In 1923 O. Perron published a method for solving the Dirichlet boundary value problem

$$\begin{cases} \Delta h = 0 \text{ in } \Omega, \\ h = g \text{ on } \partial\Omega \end{cases}$$

and it is of interest, especially if $\partial\Omega$ or g are irregular. The same method works with virtually no essential modifications for many other partial differential equations obeying a comparison principle. We will treat it for the p-Laplace equation. The p-superharmonic and p-subharmonic functions are the building blocks. This chapter is based on [GLM].

Suppose for simplicity that the domain Ω is bounded in $\mathbf{R}^n$. Let $g : \partial\Omega \to [-\infty, \infty]$ denote the desired boundary values. To begin with, g does not even have to be a measurable function. In order to solve the boundary value problem, we will construct two functions, the upper Perron solution $\overline{\mathfrak{h}}$ and the lower Perron solution $\underline{\mathfrak{h}}$. Always, $\underline{\mathfrak{h}} \le \overline{\mathfrak{h}}$ and the situation $\underline{\mathfrak{h}} = \overline{\mathfrak{h}}$ is important; in this case we write $\mathfrak{h}$ for the common function $\underline{\mathfrak{h}} = \overline{\mathfrak{h}}$.

These functions have the following properties:

(1) $\underline{\mathfrak{h}} \le \overline{\mathfrak{h}}$ in Ω
(2) $\underline{\mathfrak{h}}$ and $\overline{\mathfrak{h}}$ are p-harmonic functions, if they are finite
(3) $\underline{\mathfrak{h}} = \overline{\mathfrak{h}}$, if g is continuous
(4) If there exists a p-harmonic function h in Ω such that

$$\lim_{x \to \xi} h(x) = g(\xi)$$

at each $\xi \in \partial\Omega$, then $h = \underline{\mathfrak{h}} = \overline{\mathfrak{h}}$.
(5) If, in addition, $g \in W^{1,p}(\Omega)$ and if h is the p-harmonic function with Sobolev boundary values $h - g \in W_0^{1,p}(\Omega)$, then $h = \underline{\mathfrak{h}} = \overline{\mathfrak{h}}$.

© The Author(s), under exclusive license to Springer Nature Switzerland AG 2019
P. Lindqvist, *Notes on the Stationary p-Laplace Equation*,
SpringerBriefs in Mathematics, https://doi.org/10.1007/978-3-030-14501-9_6

There are more properties to list, but we stop here. Notice that (5) indicates that the Perron method is more general than the variational method, which uses a Hilbert space when $p = 2$.

We begin the construction by defining two classes of functions: the upper class $\mathfrak{U}_g$ and the lower class $\mathfrak{L}_g$. The *upper class* $\mathfrak{U}_g$ consists of all functions $v : \Omega \to (-\infty, \infty]$ such that

   (i) v is p-superharmonic in Ω,
   (ii) v is bounded below,
   (iii) $\liminf\limits_{x \to \xi} v(x) \geq g(\xi)$, when $\xi \in \partial\Omega$.

The *lower class* $\mathfrak{L}_g$ has a symmetric definition. We say that $u \in \mathfrak{L}_g$ if

   (i) u is p-subharmonic in Ω,
   (ii) u is bounded above,
   (iii) $\limsup\limits_{x \to \xi} u(x) \leq g(\xi)$, when $\xi \in \partial\Omega$.

It is a temptation to replace the third condition by $\lim v(x) = g(\xi)$, but that does not work. Neither is the requirement $\liminf v(x) = g(\xi)$ a good one. The reason is that we must be able to guarantee that the class is non-empty.

Notice that if $v_1, v_2, \ldots, v_k \in \mathfrak{U}_g$, then also the pointwise minimum

$$\min\{v_1, v_2, \ldots, v_k\}$$

belongs to $\mathfrak{U}_g$. (A corresponding statement about $\max\{u_1, u_2, \ldots, u_k\}$ holds for $\mathfrak{L}_g$.) This is one of the main reasons for not assuming any differentiability of p-superharmonic functions in their definition. (However, when it comes to Perron's method it does no harm to assume continuity.) It is important that the Poisson modification is possible: if v belongs to the class $\mathfrak{U}_g$, so does its Poisson modification V; recall subsection 5.4.

After these preliminaries we define at each point in Ω

$$\text{the } upper solution \ \overline{\mathfrak{h}}_g(x) = \inf_{v \in \mathfrak{U}_g} v(x),$$

$$\text{the } lower solution \ \underline{\mathfrak{h}}_g(x) = \sup_{u \in \mathfrak{L}_g} u(x).$$

Often, the subscript g is omitted. Thus we write $\overline{\mathfrak{h}}$ for $\overline{\mathfrak{h}}_g$. Before going further, let us examine an example for Laplace's equation.

Example Let Ω denote the punctured unit disc $0 < r < 1, r = \sqrt{x^2 + y^2}$, in the xy-plane. The boundary consists of a circle and a point (the origin). We prescribe the (continuous) boundary values

$$g(0, 0) = 1; \quad g = 0 \text{ when } r = 1.$$

We have

$$0 \leq \underline{\mathfrak{h}}(x, y) \leq \overline{\mathfrak{h}}(x, y) \leq \varepsilon \log(1/\sqrt{x^2 + y^2})$$

for $\varepsilon > 0$, because $0 \in \mathcal{L}_g$ and $\varepsilon \log(1/r) \in \mathfrak{U}_g$ and always $\underline{\mathfrak{h}} \leq \overline{\mathfrak{h}}$. Letting $\varepsilon \to 0$, we obtain

$$\underline{\mathfrak{h}} = \overline{\mathfrak{h}} = 0.$$

Although the Perron solutions coincide, they take the wrong boundary values at the origin! In fact, the harmonic function sought for does not exists, not with boundary values in the classical sense. However, $\mathfrak{h} - g \in W_0^{1,2}(\Omega)$ if g is smoothly extended.

A similar reasoning applies to the p-Laplace equation in a punctured ball in $\mathbf{R}^n$, when $1 < p \leq n$. – In the case $p > n$ the solution is

$$1 - |x|^{(p-n)/(p-1)}$$

and now it attains the right boundary values.

The next theorem is fundamental.

Theorem 6.1 *The function* $\overline{\mathfrak{h}}$ *satisfies one of the conditions:*

(i) $\overline{\mathfrak{h}}$ *is p-harmonic in* Ω,
(ii) $\overline{\mathfrak{h}} \equiv \infty$ *in* Ω,
(iii) $\overline{\mathfrak{h}} \equiv -\infty$ *in* Ω.

A similar result holds for $\underline{\mathfrak{h}}$.

The cases (ii) and (iii) require a lot of pedantic attention in the proof. For a succinct presentation we assume from now on that

$$m \leq g(\xi) \leq M, \quad \text{when } \xi \in \partial\Omega. \tag{6.2}$$

Now the constants m and M belong to $\mathcal{L}_g$ and $\mathfrak{U}_g$ respectively. Thus $m \leq \underline{\mathfrak{h}} \leq \overline{\mathfrak{h}} \leq M$. If $v \in \mathfrak{U}_g$, so does the cut function $\min\{v, M\}$. Cutting off all functions, we may assume that every function in sight takes values only in the interval $[m, M]$. The proof of the theorem relies on a lemma.

Lemma 6.3 *If* g *is bounded,* $\overline{\mathfrak{h}}_g$ *and* $\underline{\mathfrak{h}}_g$ *are continuous in* Ω.

Proof. Let $x_0 \in \Omega$ and $B(x_0, R) \subset\subset \Omega$. Given $\varepsilon > 0$ we will find a radius $r > 0$ such that

$$|\overline{\mathfrak{h}}(x_1) - \overline{\mathfrak{h}}(x_2)| < 2\varepsilon \quad \text{when } x_1, x_2 \in B(x_0, r).$$

Suppose that $x_1, x_2 \in B(x_0, r)$. We can find functions $v_i \in \mathfrak{U}$ such that

$$\lim_{i \to \infty} v_i(x_1) = \overline{\mathfrak{h}}(x_1), \quad \lim_{i \to \infty} v_i(x_2) = \overline{\mathfrak{h}}(x_2).$$

Indeed, if $v_i^1(x_1) \to \overline{\mathfrak{h}}(x_1)$ and $v_i^2(x_2) \to \overline{\mathfrak{h}}(x_2)$ we can use $v_i = \min\{v_i^1, v_i^2\}$. Consider the Poisson modifications

$$V_i = P(v_i, B(x_0, R)).$$

It is decisive that $V_i \in \mathfrak{U}$. By Proposition 5.13 $\overline{\mathfrak{h}} \le V_i \le v_i$ in Ω. Take i so large that

$$v_i(x_1) < \overline{\mathfrak{h}}(x_1) + \varepsilon, \ v_i(x_2) < \overline{\mathfrak{h}}(x_2) + \varepsilon.$$

It follows that

$$\overline{\mathfrak{h}}(x_2) - \overline{\mathfrak{h}}(x_1) < V_i(x_2) - V_i(x_1) + \varepsilon$$
$$\le \operatorname*{osc}_{B(x_0,r)} V_i + \varepsilon.$$

Recall that V_i is p-harmonic in $B(x_0, R)$. The Hölder continuity (Theorem 2.19) yields

$$\operatorname*{osc}_{B(x_0,r)} V_i \le L\Big(\frac{r}{R}\Big)^\alpha \operatorname*{osc}_{B(x_0,R)} V_i \le L\Big(\frac{r}{R}\Big)^\alpha (M - m)$$

when $0 < r < R/2$. Thus

$$\overline{\mathfrak{h}}(x_2) - \overline{\mathfrak{h}}(x_1) < \varepsilon + \varepsilon = 2\varepsilon$$

when r is small enough. By symmetry, $\overline{\mathfrak{h}}(x_1) - \overline{\mathfrak{h}}(x_2) < 2\varepsilon$. The continuity of $\overline{\mathfrak{h}}$ follows.

A similar proof goes for $\underline{\mathfrak{h}}$.

Proof. of Theorem 6.1 We claim that $\overline{\mathfrak{h}}$ is a solution, having assumed (6.2) for simplicity. Let $q_1, q_2, \ldots, q_\nu, \ldots$ be the rational points in Ω. We will first construct a sequence of functions in the upper class $\mathfrak{U}$ converging to $\overline{\mathfrak{h}}$ at the rational points. Given q_ν we can find $v_1^\nu, v_2^\nu, \ldots$ in $\mathfrak{U}$ such that

$$\overline{\mathfrak{h}}(q_\nu) \le v_i^\nu(q_\nu) < \overline{\mathfrak{h}}(q_\nu) + \frac{1}{i}, \quad i = 1, 2, 3, \ldots.$$

Define
$$w_i = \min\{v_1^1, v_2^1, \ldots, v_i^1, v_1^2, v_2^2, \ldots, v_i^2, \ldots, v_1^i, v_2^i, \ldots, v_i^i\}$$

Then $w_i \in \mathfrak{U}$, $w_1 \ge w_2 \ge \ldots$ and

$$\overline{\mathfrak{h}}(q_\nu) \le w_i(q_\nu) \le v_i^\nu(q_\nu) \quad \text{when } i \ge \nu.$$

Hence $\lim w_i(q_\nu) = \overline{\mathfrak{h}}(q_\nu)$ at each rational point, as desired.

Suppose that $B \subset\subset \Omega$ and consider the Poisson modification

$$W_i = P(w_i, B).$$

Since also $W_i \in \mathfrak{U}$, we have

$$\overline{\mathfrak{h}} \leq W_i \leq w_i.$$

Thus $\lim W_i(q_\nu) = \overline{\mathfrak{h}}(q_\nu)$ at the rational points. In other words, W_i is better than w_i. We also conclude that $W_1 \geq W_2 \geq W_3 \geq \ldots$. According to Harnack's convergence theorem (Theorem 5.10)

$$W = \lim_{i \to \infty} W_i$$

is p-harmonic in B. By the construction $W \geq \overline{\mathfrak{h}}$ and $W(q_\nu) = \overline{\mathfrak{h}}(q_\nu)$ at the rational points. We have two continuous functions, the p-harmonic W and $\overline{\mathfrak{h}}$ (Lemma 6.3), that coincide in a dense subset. Then they coincide everywhere. The conclusion is that in B we have $\overline{\mathfrak{h}} =$ the p-harmonic function W. Thus $\overline{\mathfrak{h}}$ is p-harmonic in B. It follows that $\overline{\mathfrak{h}}$ is p-harmonic also in Ω.

A similar proof applies to $\underline{\mathfrak{h}}$.

We have learned that the Perron solutions are p-harmonic functions, if they take finite values. Always

$$-\infty \leq \underline{\mathfrak{h}} \leq \overline{\mathfrak{h}} \leq \infty$$

but the situation $\overline{\mathfrak{h}} \neq \underline{\mathfrak{h}}$ is possible. When $\underline{\mathfrak{h}} = \overline{\mathfrak{h}}$ we denote the common function with $\mathfrak{h}$.

Theorem 6.4 (Wiener's resolutivity theorem) *Suppose that $g : \partial\Omega \to \mathbf{R}$ is continuous. Then $\underline{\mathfrak{h}}_g = \overline{\mathfrak{h}}_g$ in Ω.*

Proof. Our proof is taken from [LM]. For the proof we need to know that there is an exhaustion of Ω with *regular* domains $D_j \subset\subset \Omega$,

$$\Omega = \bigcup_{j=1}^{\infty} D_j, \quad D_1 \subset D_2 \subset \ldots$$

The domain D_j can be constructed as a union of cubes or as a domain with a smooth boundary.

We first do a reduction. If we can prove the theorem for smooth g's, we are done. Indeed, given $\varepsilon > 0$ there is a smooth φ such that

$$\varphi(\xi) - \varepsilon < g(\xi) < \varphi(\xi) + \varepsilon, \quad \text{when } \xi \in \partial\Omega.$$

Thus,

$$\mathfrak{h}_\varphi - \varepsilon \leq \mathfrak{h}_{\varphi-\varepsilon} \leq \underline{\mathfrak{h}}_g \leq \overline{\mathfrak{h}}_g \leq \mathfrak{h}_{\varphi+\varepsilon} = \mathfrak{h}_\varphi + \varepsilon,$$

if $\underline{\mathfrak{h}}_\varphi = \overline{\mathfrak{h}}_\varphi$. Since $\varepsilon > 0$ was arbitrary, we conclude that $\underline{\mathfrak{h}}_g = \overline{\mathfrak{h}}_g$. Thus we can assume that $g \in C^\infty(\mathbf{R}^n)$. What we need is only $g \in W^{1,p}(\Omega) \cap C(\overline{\Omega})$.

The proof, after the reduction to the situation $g \in C^\infty(\mathbf{R}^n)$, relies on the unique-ness of the solution to the Dirichlet problem with boundary values in Sobolev's sense. In virtue of Theorem 2.16 there is a *unique p-harmonic function* $h \in C(\Omega) \cap W^{1,p}(\Omega)$ in Ω with boundary values $h - g \in W_0^{1,p}(\Omega)$. Nothing has to be assumed about the domain Ω, except that it is bounded, of course. We claim that $h \geq \overline{\mathfrak{h}}$ and $h \leq \underline{\mathfrak{h}}$, which implies the desired resolutivity $\underline{\mathfrak{h}} = \overline{\mathfrak{h}}$. To this end, let v denote the solution to the obstacle problem with g acting as obstacle. See Theorem 5.6. Then $v - g \in W_0^{1,p}(\Omega)$ and $v \geq g$ in Ω. Since v is a weak supersolution, $v \in \mathfrak{U}_g$. (The reason for introducing the auxiliary function v is that one cannot guarantee that h itself belongs to the upper class! However, the obstacle causes $v \geq g$.)

Construct the sequence of Poisson modifications

$$V_1 = P(v, D_1), \quad V_2 = P(v, D_2) = P(V_1, D_2),$$
$$V_3 = P(v, D_3) = P(V_2, D_3), \dots$$

Then $V_1 \geq V_2 \geq V_3 \geq \dots$ and $V_j \in \mathfrak{U}_g$. Also $v_j - g \in W_0^{1,p}(\Omega)$ and

$$\int_\Omega |\nabla V_j|^p dx \leq \int_\Omega |\nabla v|^p dx \leq \int_\Omega |\nabla g|^p dx. \tag{6.5}$$

We have $\overline{\mathfrak{h}}_g \leq V_j$. Using Harnack's convergence theorem (Theorem 5.10) we see that

$$V = \lim_{j \to \infty} V_j$$

is p-harmonic in D_1, in $D_2, \dots$, and hence in Ω. But (6.5) and the fact that $V_j - g \in W_0^{1,p}(\Omega)$ show that $V - g \in W_0^{1,p}(\Omega)$. Thus V solves the same problem as h. The aforementioned uniqueness implies that $V = h$ in Ω.

We have obtained the result

$$\overline{\mathfrak{h}}_g \leq \lim V_j = h,$$

as desired. The inequality $\underline{\mathfrak{h}}_g \geq h$ has a similar proof. The theorem follows. □

As a byproduct of the proof we obtain the following.

Proposition 6.6 *If $g \in W^{1,p}(\Omega) \cap C(\overline{\Omega})$, then the p-harmonic function with bound-ary values in Sobolev's sense coincides with the Perron solution $\mathfrak{h}_g$.*

The question about at which boundary points the prescribed continuous boundary values are attained (in the classical sense) can be restated in terms of so-called barriers, a kind of auxiliary functions. Let Ω be a bounded domain. We say that $\xi \in \partial\Omega$ is a *regular boundary point*, if

$$\lim_{x \to \xi} \mathfrak{h}_g(x) = g(\xi)$$

for all $g \in C(\partial\Omega)$.

Remark There is an equivalent definition of a regular boundary point ξ. The equation

$$\Delta_p u = -1$$

has a unique weak solution $u \in W_0^{1,p}(\Omega) \cap C(\Omega)$. The point ξ is regular if and only if

$$\lim_{x \to \xi} u(x) = 0.$$

The advantage is that only one function is involved. The proof of the equivalence of the definitions is difficult.

Definition 6.7 A point $\xi_0 \in \partial\Omega$ has a *barrier* if there exists a function $w : \Omega \to \mathbf{R}$ such that

  (i) w is p-superharmonic in Ω,
 (ii) $\liminf_{x \to \xi} w(x) > 0$ for all $\xi \neq \xi_0, \xi \in \partial\Omega$,
(iii) $\lim_{x \to \xi_0} w(x) = 0$.

The function w is called a barrier.

Theorem 6.8 *Let Ω be a bounded domain. The point $\xi_0 \in \partial\Omega$ is regular if and only if there exists a barrier at ξ_0.*

Proof. The proof of that the existence of a barrier is sufficient for regularity is completely analogous to the classical proof. Let $\varepsilon > 0$ and $M = \sup |g|$. We can use the assumptions to find $\delta > 0$ and $\lambda > 0$ such that

$$
\begin{aligned}
|g(\xi) - g(\xi_0)| &< \varepsilon, & \text{when } |\xi - \xi_0| &< \delta; \\
\lambda w(x) &\geq 2M, & \text{when } |x - \xi_0| &\geq \delta.
\end{aligned}
$$

This has the consequence that the functions $g(\xi_0) + \varepsilon + \lambda w(x)$ and $g(\xi_0) - \varepsilon - \lambda w(x)$ belong to the classes $\mathfrak{U}_g$ and $\mathfrak{L}_g$, respectively. Thus

$$g(\xi_0) - \varepsilon - \lambda w(x) \leq \underline{\mathfrak{h}}_g(x) \leq \overline{\mathfrak{h}}_g(x) \leq g(\xi_0) + \varepsilon + \lambda w(x)$$

or

$$|\mathfrak{h}_g(x) - g(\xi_0)| \leq \varepsilon + \lambda w(x)$$

Since $w(x) \to 0$ as $x \to \xi_0$, we obtain that $\mathfrak{h}_g(x) \to g(\xi_0)$ as $x \to \xi_0$. Thus ξ_0 is a regular boundary point.

For the opposite direction we assume that ξ_0 is regular. In order to construct the barrier we take

$$g(x) = |x - \xi_0|^{\frac{p}{p-1}}.$$

An easy calculation shows that $\Delta_p g(x)$ is a positive constant, when $x \neq \xi_0$. We conclude that g is p-subharmonic in Ω. Let $\overline{\mathfrak{h}}_g$ denote the corresponding upper Perron solution, when g are the boundary values. By the comparison principle $\overline{\mathfrak{h}}_g \geq g$ in Ω. Because ξ_0 is assumed to be a regular boundary point, we have

$$\lim_{x \to \xi_0} \overline{\mathfrak{h}}_g(x) = g(\xi_0) = 0.$$

Hence $w = \overline{\mathfrak{h}}_g$ will do as a barrier.   □

Example $1 < p \leq n$. Suppose that Ω satisfies the well-known *exterior sphere condition*. Then each boundary point is regular. For the construction of a barrier at $\xi_0 \in \partial\Omega$ we assume that $\overline{B(x_0, R)} \cap \overline{\Omega} = \{\xi_0\}$. The function

$$w(x) = \int_R^{|x-x_0|} t^{-\frac{n-1}{p-1}} dt$$

will do as a barrier.

Example $p > n$. Without any hypothesis

$$w(x) = |x - \xi_0|^{\frac{p-n}{p-1}}$$

will do as a barrier at ξ_0. Thus every boundary point of an arbitrary domain is regular, when $p > n$.

An immediate consequence of Theorem 6.6 is the following result, indicating that *the more complement the domain has, the better the regularity is.* If $\Omega_1 \subset \Omega_2$ and if $\xi_0 \in \partial\Omega_1 \cap \partial\Omega_2$, then, if ξ_0 is regular with respect to Ω_2, so it is with respect to Ω_1. The reason is that the barrier for Ω_2 is a barrier for Ω_1.

The concept of a barrier is rather implicit in a general situation. A much more advanced characterization of the regular boundary points is the celebrated Wiener criterion, originally formulated for the Laplace equation in 1924 by N. Wiener. He used the electrostatic capacity. We need the p-capacity.

The *p-capacity* of a closed set $E \subset\subset B_r$ is defined as

$$\mathrm{Cap}_p(E, B_r) = \inf_\zeta \int_{B_r} |\nabla\zeta|^p dx$$

where $\zeta \in C_0^\infty(B_r)$, $0 \leq \zeta \leq 1$ and $\zeta = 1$ in E. The *Wiener criterion* can now be stated.

Theorem 6.9 *The point $\xi_0 \in \partial\Omega$ is regular if and only if the integral*

$$\int\limits_0^1 \left[\frac{\text{Cap}_p(\overline{B(\xi_0, t)} \cap E, B(\xi_0, 2t))}{\text{Cap}_p(\overline{B(\xi_o, t)}, B(\xi_0, 2t))} \right]^{\frac{1}{p-1}} \frac{dt}{t} = \infty$$

diverges, where $E = \mathbf{R}^n \setminus \Omega$.

The Wiener criterion with p was formulated in 1970 by V. Mazja. He proved the sufficiency, [Ma]. For the necessity, see [KM2]. The case $p > n - 1$ has a simpler proof, written down for $p = n$ in [LM]. The proofs are too difficult to be given here.

One can say the following when p varies but the domain is kept fixed. The greater p is, the better for regularity. *If ξ_0 is p_1-regular, then ξ_0 is p_2-regular for all $p_2 \geq p_1$.* This deep result can be extracted from the Wiener criterion. The Wiener criterion is also the fundament for the so-called *Kellogg property*: The irregular boundary points of a given domain form a set of zero p-capacity. Roughly speaking, this means that the huge majority of the boundary points is regular.

It would be nice to find simpler proofs when it comes to the Wiener criterion!

Chapter 7
Some Remarks in the Complex Plane

For elliptic partial differential equations it is often the case that in only two variables the theory is much richer than in higher dimensions. Indeed, also the p-Laplace equation

$$\frac{\partial}{\partial x}\left(|\nabla u|^{p-2}\frac{\partial u}{\partial x}\right) + \frac{\partial}{\partial y}\left(|\nabla u|^{p-2}\frac{\partial u}{\partial y}\right) = 0 \tag{7.1}$$

in two variables, x and y, exhibits an interesting structure, not known of in space. It lives a life of its own in the plane! Among other things a remarkable generalization of the Cauchy-Riemann equations is possible. The hodograph method can be used to obtain many explicit solutions.

In the plane the advanced theory of quasiconformal mappings is available for equations of the type

$$\frac{\partial}{\partial x}\left(\varrho(|\nabla u|)\frac{\partial u}{\partial x}\right) + \frac{\partial}{\partial y}\left(\varrho(|\nabla u|)\frac{\partial u}{\partial y}\right) = 0$$

and is described in the book "Mathematical Aspects of Subsonic and Transonic Gas Dynamics" by L. Bers. (The flow is subsonic, if the Mach number

$$M^2 = -\frac{q\,\varrho'(q)}{\varrho(q)} \qquad q = |\nabla u|$$

is in the range $0 \le M^2 < 1$. The quasiregular mappings work if the Mach number is bounded away from $-\infty$ and 1, but it can be negative and is $2 - p$ in our case). The p-Laplace equation presents some difficulties at the critical points ($\nabla u = 0$). It was shown by B. Bojarski and T. Iwaniec in 1983 that

$$f = \frac{\partial u}{\partial x} - i\frac{\partial u}{\partial y} \qquad (i^2 = -1)$$

© The Author(s), under exclusive license to Springer Nature Switzerland AG 2019
P. Lindqvist, *Notes on the Stationary p-Laplace Equation*,
SpringerBriefs in Mathematics, https://doi.org/10.1007/978-3-030-14501-9_7

is a quasiregular mapping (=quasiconformal, except injective). The most important consequence is that the zeros of f, that is, the critical points of the p-harmonic function u, are isolated. Thus they are points, as the name suggests.

Theorem 7.2 (Bojarski-Iwaniec) *Let u be a p-harmonic function in the domain Ω in the plane. Then the complex gradient $f = u_x - iu_y$ is a quasiregular mapping, that is:*

(i) *f is continuous in Ω*
(ii) *$u_x, u_y \in W_{loc}^{1,2}(\Omega)$*
(iii) *$\left| \frac{\partial f}{\partial \bar{z}} \right| \le \left| 1 - \frac{2}{p} \right| \left| \frac{\partial f}{\partial z} \right|$ a.e. in Ω.*

Remark It is essential that $|1 - 2/p| < 1$. The notation

$$\frac{\partial}{\partial z} = \frac{1}{2}\left(\frac{\partial}{\partial x} - i\frac{\partial}{\partial y} \right), \quad \frac{\partial}{\partial \bar{z}} = \frac{1}{2}\left(\frac{\partial}{\partial x} + i\frac{\partial}{\partial y} \right)$$

is convenient. The proof is given in [BI1] where also the formula

$$\frac{\partial f}{\partial \bar{z}} = \left(\frac{1}{p} - \frac{1}{2} \right)\left(\frac{\bar{f}}{f}\frac{\partial f}{\partial z} + \frac{f}{\bar{f}}\overline{\left(\frac{\partial f}{\partial z} \right)} \right)$$

is established.

By the general theory, the zeros of a quasiregular mapping are isolated, except when the mapping is identically zero. We infer that *the critical points*

$$S = \{(x, y)|\ \nabla u(x, y) = 0\}$$

of a p-harmonic function u are isolated, except when the function is a constant. Outside the set S the function is real-analytic. According to a theory due to Y. Reshetnyak an elliptic partial differential equation is associated to a quasiregular mapping, cf [Re] and [BI2]. In the plane this equation is always a linear one. In the present case u_x, u_y and $\log |\nabla u|$ are solutions to the same linear equation. However, this equation depends on ∇u itself! A different approach to find an equation for $\log |\nabla u|$ has been suggested by Alessandrini, cf [Al].

Next, let us consider a counterpart to the celebrated *Cauchy-Riemann equations*. If u is p-harmonic in a simply connected domain Ω, then there is a function v, unique up to a constant, such that

$$v_x = -|\nabla u|^{p-2}u_y, \quad v_y = |\nabla u|^{p-2}u_x$$

or, equivalently,

$$u_x = |\nabla v|^{q-2}v_y, \quad u_y = -|\nabla v|^{q-2}v_x$$

in Ω. For smooth functions this is evident from (7.1) but the general case is harder. In particular, $|\nabla u|^p = |\nabla v|^q$ and $1/p + 1/q = 1$. *The conjugate function v is q-harmonic in Ω, q being the conjugate exponent:*

$$\frac{1}{p} + \frac{1}{q} = 1.$$

A most interesting property is that

$$\langle \nabla u, \nabla v \rangle = 0.$$

Therefore *the level curves of u and v are orthogonal to each other*, apart from the singular set S. A good example is

$$u + iv = \frac{p-1}{p-2}\left(|z|^{\frac{p-2}{p-1}} - 1\right) + i \arg z,$$

where $z = x + iy$ and Ω is the complex plane (with a slit from 0 to ∞). We refer to [AL] for this kind of function theory.

A lot of explicit examples are given in [A5]. The optimal regularity of a p-harmonic function in the plane has been determined by T. Iwaniec and J. Manfredi.

Theorem 7.3 (Iwaniec-Manfredi) *Every p-harmonic function, $p \neq 2$, is of class $C^{k,\alpha}_{loc}(\Omega) \cap W^{k+2,q}_{loc}(\Omega)$, where the integer $k > 1$ and the exponent α, $0 < \alpha \leq 1$, are determined by the formula*

$$k + \alpha = 1 + \frac{1}{6}\left(1 + \frac{1}{p-1} + \sqrt{1 + \frac{14}{p-1} + \frac{1}{(p-1)^2}}\right).$$

The summability exponent q is any number in the range

$$1 \leq q < \frac{2}{2-\alpha}.$$

Proof The proof is based on a hodograph representation, see [IM]. □

Remark (1) Notice that always $u \in W^{3,1}_{loc}(\Omega)$. Therefore u has Sobolev derivatives of order three.
(2) As $p \to \infty$, the above formula does not produce the correct regularity class for the limit equation. The reason is subtle.
(3) As $p \to 1, k \to \infty$. However "1-harmonic functions" are not of class C^∞.

There are several properties that have been established in the plane but, so far as we know, not in space. A few of them are:

The Principle of Unique Continuation. Suppose that u is a p-harmonic function in Ω and that $u \equiv 0$ in a ball $B \subset \Omega$. Then $u \equiv 0$ in Ω.
The Strong Comparison Principle. Suppose that u and v are p-harmonic functions and that $u \leq v$ in Ω. If $u(x_0) = v(x_0)$ at some point $x_0 \in \Omega$, then $u \equiv v$ in Ω. – For a proof we refer to [M1].

Chapter 8
The Infinity Laplacian

The limit equation of the p-Laplace equation as $p \to \infty$ is a very fascinating one. In two variables it is the equation

$$u_x^2 u_{xx} + 2u_x u_y u_{xy} + u_y^2 u_{yy} = 0,$$

which was found in 1967 by G. Aronsson, cf [A1]. It provides the best Lipschitz extension of given boundary values and has applications in image processing. It requires the modern concept of *viscosity solutions*, originally developed for equations of the first order (Hamilton-Jacobi equations). The equation has a connection to Stochastic Game Theory.

The ∞-Laplace operator

$$\Delta_\infty u = \sum_{i,j=1}^{n} \frac{\partial u}{\partial x_i} \frac{\partial u}{\partial x_j} \frac{\partial^2 u}{\partial x_i \partial x_j} = \frac{1}{2}\langle \nabla u, \nabla |\nabla u|^2 \rangle$$

comes from the following consideration. Start with

$$\Delta_p u = |\nabla u|^{p-4}\left\{|\nabla u|^2 \Delta u + (p-2)\Delta_\infty u\right\} = 0,$$

divide out the factor $|\nabla u|^{p-4}$, and let $p \to \infty$ in

$$\frac{|\nabla u|^2 \Delta u}{p-2} + \Delta_\infty u = 0.$$

This leads to the equation

$$\Delta_\infty u = 0.$$

© The Author(s), under exclusive license to Springer Nature Switzerland AG 2019
P. Lindqvist, *Notes on the Stationary p-Laplace Equation*,
SpringerBriefs in Mathematics, https://doi.org/10.1007/978-3-030-14501-9_8

However, this derivation of the ∞-Laplace equation leaves much to be desired. Nevertheless, the equation is the correct one.

For a finite p the equation $\Delta_p u = 0$ is the Euler-Lagrange equation for the variational integral

$$\|\nabla u\|_p = \left\{ \int_\Omega |\nabla u|^p dx \right\}^{\frac{1}{p}}.$$

Hence one may expect the equation $\Delta_\infty u = 0$ to be the Euler-Lagrange equation for the "functional"

$$\|\nabla u\|_\infty = \lim_{p \to \infty} \|\nabla u\|_p = \operatorname{ess\,sup}_{x \in \Omega} |\nabla u(x)|.$$

Thus the minimax problem

$$\min_u \max_x |\nabla u(x)|$$

is, as it were, involved here. All this can be done rigorously.

The equation has an interesting *geometric interpretation*, though valid only for rather smooth functions. To explain it via the "gradient flow" we consider the curve $x = x(t) = (x_1(t), \ldots, x_n(t))$ in $\mathbf{R}^n$. Follow $|\nabla u|^2$ along the curve. Differentiating $|\nabla u(x(t))|^2$ we obtain

$$\frac{d}{dt} |\nabla u|^2 = 2 \sum \frac{\partial u}{\partial x_i} \frac{\partial^2 u}{\partial x_i \partial x_j} \frac{dx_j}{dt}.$$

We observe that if the curve is a solution of the dynamical system (the so-called gradient flow)

$$\frac{dx}{dt} = \nabla u(x(t))$$

we obtain, replacing $\frac{dx_j}{dt}$ by $\frac{\partial u}{\partial x_j}$,

$$\frac{d}{dt} |\nabla u|^2 = 2 \Delta_\infty u$$

taken along the curve. So far, u is arbitrary. Thus, if the original u was a solution of $\Delta_\infty u = 0$, we conclude that $|\nabla u|$ is a constant along the curve. Since ∇u represents the normal direction to the level surfaces of u, we have the following interpretation. *Along a stream line $|\nabla u|$ is constant.* However, different stream lines usually have different constants. This property is useful for applications to image processing.

The ∞-Laplacian also appears in an amusing formula. In the Taylor expansion

$$u(x + h) = u(x) + \langle \nabla u(x), h \rangle + \frac{1}{2} \langle h, D^2 u(x) h \rangle + \ldots$$

we take $h = t \nabla u(x)$. We arrive at

$$u(x + t\nabla u(x)) = u(x) + t|\nabla u(x)|^2 + \frac{1}{2}t^2\Delta_\infty u(x) + \dots$$

to our pleasure. The t^2-term contains the ∞-Laplacian. The resulting formula

$$\frac{u(x + t\nabla u(x)) - 2u(x) + u(x - t\nabla u(x))}{t^2} = \Delta_\infty u(x) + \dots$$

can be utilized in a numerical scheme.

A few explicit solutions are

$$a\sqrt{x_1^2 + \cdots + x_k^2} + b \quad (1 \le k \le n)$$
$$a_1 x_1 + \cdots + a_n x_n + b$$
$$a_1 x_1^{4/3} + \cdots + a_n x_n^{4/3} \quad \left(\sum a_j^3 = 0\right)$$

as well as all angles in spherical coordinates like

$$\arctan\left(\frac{x_2}{x_1}\right), \quad \arctan\left(\frac{x_3}{\sqrt{x_1^2 + x_2^2}}\right).$$

Expressions in disjoint variables like

$$5\sqrt{x_1^2 + x_2^2} + 3\sqrt{x_3^2 + x_4^2} + (x_5^{4/3} - x_6^{4/3})$$

can be added to the list.[1] Finally, we mention the solutions of the *eikonal equation* $|\nabla u|^2 = 1$. – Many examples in two variables are constructed in [A3].

The Dirichlet problem is to find a solution to

$$\begin{cases} \Delta_\infty u = 0 \text{ in } \Omega, \\ u = g \text{ on } \partial\Omega \end{cases}$$

in a bounded domain Ω. (In two variables the equation is formally classified as a parabolic one but the boundary values are prescribed as for elliptic equations!) The difficulty here is the concept of solutions, because u is not always of class C^2. We will return to the concept of solutions later. Suppose now that $g : \partial\Omega \to \mathbf{R}$ is a Lipschitz continuous function, that is

$$|g(\xi_1) - g(\xi_2)| \le L|\xi_1 - \xi_2|$$

when $\xi_1, \xi_2 \in \partial\Omega$. We may extend g to be defined in Ω using one of the formulas

---

[1] For *viscosity* solutions some care is called for, see [HF].

$$g(x) = \max_{\xi \in \partial\Omega} \big(g(\xi) - L|x - \xi|\big) \text{ or } g(x) = \min_{\xi \in \partial\Omega} \big(g(\xi) + L|x - \xi|\big).$$

The extended function has the same Lipschitz constant L. By Rademacher's theorem ∇g exists a.e. and $|\nabla g| \le L$. Therefore we may assume that $g \in C(\overline{\Omega}) \cap W^{1,\infty}(\Omega)$. Now we want to construct the solution by letting $p \to \infty$. Let $p > n$. As we know, there is a unique p-harmonic function $u_p \in C(\overline{\Omega}) \cap W^{1,p}(\Omega)$ such that $u_p = g$ on $\partial\Omega$. (Since $p > n$, the regularity of Ω plays no role now.) We have

$$\left\{ \fint_\Omega |\nabla u_p|^s dx \right\}^{\frac{1}{s}} \le \left\{ \fint_\Omega |\nabla u_p|^p dx \right\}^{\frac{1}{p}}$$

$$\le \left\{ \fint |\nabla g|^p dx \right\}^{\frac{1}{p}} \le |\Omega|^{-\frac{1}{p}} L$$

as soon as $p > s$. Using some compactness arguments we can conclude the following.

Proposition 8.1 *There is a subsequence u_{p_k} and a function $u_\infty \in C(\overline{\Omega}) \cap W^{1,\infty}(\Omega)$ such that $u_{p_k} \to u_\infty$ uniformly in Ω and $\nabla u_{p_k} \rightharpoonup \nabla u_\infty$ weakly in each $L^s(\Omega)$. In particular, $u_\infty = g$ on $\partial\Omega$.*

The so obtained u_∞ is called a *variational solution* of the equation. Several questions arise. Is u_∞ unique or does it depend on the particular subsequence chosen? How is u_∞ related to the limit equation $\Delta_\infty u = 0$? Is it a "solution"? At least it follows directly from the construction that u_∞ has a minimizing property:

Lemma 8.2 *If $D \subset \Omega$ is a subdomain and if $v \in C(\overline{D}) \cap W^{1,\infty}(D)$ has boundary values $v = u_\infty$ on ∂D, then*

$$\|\nabla u_\infty\|_{L^\infty(D)} \le \|\nabla v\|_{L^\infty(D)} .$$

In view of the mean value theorem in the differential calculus, the lemma says that the Lipschitz constant of u_∞ cannot be locally improved. It is the best one.

Let us discuss the concept of solutions. In two variables the theorem below easily enables one to conclude that there are "solutions" not having second continuous derivatives.

Theorem 8.3 (Aronsson) *Suppose that $u \in C^2(\Omega)$ where Ω is a domain in $\mathbf{R}^n$. If $\Delta_\infty u = 0$ in Ω, then $\nabla u \ne 0$ in Ω, except when u reduces to a constant.*

Proof See [A2] for $n = 2$. The cases $n \ge 2$ are in [Y].   □

It turns out that the ∞-Laplace equation does not have a weak formulation with the test functions under the integral sign. Indeed, multiplying the equation with a test function and integrating leads to

$$\int_\Omega \eta \, \Delta_\infty u \, dx = 0 ,$$

an expression from which one cannot eliminate the second derivatives of u. Actually, integrations by part seem to make the situation worse!

The way out of this dead end is to use viscosity solutions as in [BDM]. It has to be written in terms of viscosity supersolutions and subsolutions.

Definition 8.4 We say that the lower semicontinuous function v is ∞-*superharmonic* in Ω, if whenever $x_0 \in \Omega$ and $\varphi \in C^2(\Omega)$ are such that

- $\varphi(x_0) = v(x_0)$
- $\varphi(x) < v(x)$, when $x \neq x_0$

then we have $\Delta_\infty \varphi(x_0) \leq 0$.

Notice that each point x_0 needs its own family of test functions (which may be empty) and that $\Delta_\infty \varphi$ is evaluated only at the point of contact. By the infinitesimal calculus $\nabla \varphi(x_0) = \nabla v(x_0)$, provided that the latter exists at all. It is known that $v \in C(\Omega)$ and that could have been incorporated in the definition.

The definition of an ∞-*subharmonic function* is similar. A function is defined to be ∞-*harmonic* if it is both ∞-superharmonic and ∞-subharmonic. Thus the ∞-harmonic functions are the viscosity solutions of the equation.

Example The function $v(x) = 1 - |x|$ is ∞-harmonic when $x \neq 0$. It is ∞-superharmonic in $\mathbf{R}^n$. At the origin there is no test function touching v from below. Thus there is no requirement to verify.

Example The interesting function

$$x^{4/3} - y^{4/3}$$

in two variables belongs to a family of solutions discovered by G. Aronsson [A3]. The reader may verify that it is ∞-harmonic, indeed. This function belongs to $C_{\mathrm{loc}}^{1,1/3}(\mathbf{R}^2)$ and to $W_{\mathrm{loc}}^{2,3/2-\varepsilon}(\mathbf{R}^2)$ for each $\varepsilon > 0$. It does not have second continuous derivatives on the coordinate axes. See also [Sa].

We have now three concepts of solutions to deal with: classical solutions, variational solutions and viscosity solutions. The inclusions

$$\{\text{classical solutions}\} \subset \{\text{variational solutions}\} \subset \{\text{viscosity solutions}\}$$

are not very difficult to prove. – In fact, all solutions are variational solutions. This follows from R. Jensen's remarkable uniqueness theorem.

Theorem 8.5 (Jensen) *Let Ω be an arbitrary bounded domain. Given a Lipschitz continuous function $g : \partial\Omega \to \mathbf{R}$ there exists one and only one viscosity solution $u_\infty \in C(\overline{\Omega})$ of the equation $\Delta_\infty u_\infty = 0$ in Ω with boundary values $u_\infty = g$ on $\partial\Omega$.*

Proof The existence is essentially Proposition 8.1. The uniqueness is proved in [J]. Jensen's uniqueness proof uses several auxiliary equations and the method of doubling the variables. Another proof is given in [BB]. In [AS] a simple proof based on comparison with cones is given.   □

We mention a characterization in terms of *comparison with cones*, cf [CEG] and [CZ]. For smooth functions the property follows from Taylor's expansion.

Theorem 8.6 (Comparison with Cones) *Let v be continuous in Ω. Then v is an ∞-superharmonic function if and only if the comparison with cones holds: if $D \subset \Omega$ is any subdomain, $a > 0$ and $x_0 \in \mathbf{R}^n \backslash D$, then $v(\xi) \leq a|\xi - x_0|$ on ∂D implies that $v(x) \leq a|x - x_0|$ in D.*

The apex x_0 is outside the domain. The ∞-harmonic functions are precisely those that obey the comparison with cones, both from above and below! This property has been used by O. Savin to prove that ∞-harmonic functions in the plane are continuously differentiable. In higher dimensions ∞-harmonic functions are, of course, differentiable at *almost* every point. In fact, they are differentiable at *every* point. The proof of this property in [ES] is of an unusual kind: no estimate giving continuity is produced.

We cannot resist mentioning that the function

$$V(x) = \int |x - y| \varrho(y) dy$$

is ∞-subharmonic for $\varrho \geq 0$, cf [CZ].[2] It is a curious coincidence that the fundamental solution of the *biharmonic equation* $\Delta \Delta u = -\delta$ in three dimensional space ($n = 3$) is $\frac{|x - x_0|}{8\pi}$ so that $\Delta \Delta V(x) = -8\pi \varrho(x)$. Given V, this tells us how to find a suitable ϱ.

Finally, we mention that the property of unique continuation does not hold. There is an example with a domain Ω and two ∞-harmonic functions u_1 and u_2 in Ω, such that $u_1 = u_2$ in an open subset of Ω but $u_1 \not\equiv u_2$ in Ω. We do not know, whether this phenomenon can occur for $u_1 \equiv 0$.

Tug-of-War. A marvellous connection between the ∞-Laplace Equation and Stochastic Game Theory was discovered in 2009 by Y. Peres, O. Schramm, S. Sheffield, and D. Wilson. A mathematical game called Tug-of-War lead to the equation. We refer directly to [PSW] about this fascinating discovery. See [Ro] for an introductory account. We shall only give a sketch,[3] describing the game without proper mathematical terms, naturally without laying claim to any exactness.

---

[2]This result due to Grandall and Zhang has the consequence that the "mysterious inequality"

$$\iiint \frac{|x - c|^2 \langle x - a, x - b \rangle - \langle x - a, x - c \rangle \langle x - b, x - c \rangle}{|x - a||x - b||x - c|^3} \varrho(a) \varrho(b) \varrho(c) da db dc \geq 0$$

has to hold for all compactly supported densities ϱ.

[3]This is taken from my lecture notes in [L4].

Recall the *Brownian Motion* for the ordinary Laplace Equation. Consider a bounded domain Ω, sufficiently regular for the Dirichlet Problem. Suppose that the boundary values

$$g(x) = \begin{cases} 1, & \text{when} \quad x \in C \\ 0, & \text{when} \quad x \in \partial\Omega \setminus C \end{cases}$$

are prescribed, where $C \subset \partial\Omega$ is closed. Let u denote the solution of $\Delta u = 0$ in Ω with boundary values g (one may take the Perron solution). If a particle starts its Brownian motion at the point $x \in \Omega$, then $u(x) =$ the probability that the particle (first) exits through C. In other words, the harmonic measure is related to Brownian motion. One may also consider more general boundary values. So much about Laplace's Equation.

Let us now describe *"Tug-of-War"*. Consider a game played by two players. A token is placed at the point $x \in \Omega$. One player tries to move it so that it leaves the domain via the boundary part C, the other one aims at the complement $\partial\Omega \setminus C$. The rules are:

- A fair coin is tossed.
- The player who wins the toss moves the token less than ε units in the most favourable direction.
- Both players play optimally.
- The game ends when the token hits the boundary $\partial\Omega$.

Let us denote

$$u_\varepsilon(x) = \text{the probability that the exit is through } C.$$

Then the 'Dynamic Programming Principle'

$$u_\varepsilon(x) = \frac{1}{2}\Big(\sup_{|e|<1} u_\varepsilon(x + \varepsilon e) + \inf_{|e|<1} u_\varepsilon(x + \varepsilon e)\Big)$$

holds. The abbreviation is $\mathbb{DPP}$. At a general point x there are usually only two optimal directions, viz. $\pm\nabla u_\varepsilon(x)/|\nabla u_\varepsilon(x)|$, one for each player. The directions are opposite, whence the name 'Tug-of-War'. The reader may consult [Ro]. As the step size ε goes to zero, one obtains a function

$$u(x) = \lim_{\varepsilon \to 0} u_\varepsilon(x).$$

The sequence converges *almost surely*. The spectacular result is that the so obtained function u is the solution to the ∞-Laplace Equation with boundary values g.

While the Brownian motion does not favour any direction, the Tug-of-War does. There is also a stochastic game for the p-harmonic Equation, though the rules are more complicated, cf. [MPR3]. See [KMP] for $p < 2$.

Chapter 9
Viscosity Solutions

The modern theory of viscosity solutions was developed by Crandall, Evans, Jensen, Ishii, Lions, and others. First, it was designed for first order equations. Later, it was extended to second order equations. The solutions must obey a Comparison Principle. We refer to [K] and [CIL] for some background.[1] For the ∞-Laplacian the concept appeared in [BDM]. The viscosity solutions of the p-Laplace Equation are the same as the p-harmonic functions, which was established in [JLM]. We shall give a simple proof of this fact, based on [JJ], though only for $p \geq 2$.

We begin with some infinitesimal calculus. Suppose that $v : \Omega \to \mathbf{R}$ is a given function. Assume that $\phi \in C^2(\Omega)$ is *touching v from below* at some point $x_0 \in \Omega$:

$$\begin{cases} \phi(x_0) = v(x_0), \\ \phi(x) < v(x) \quad \text{when} \quad x \neq x_0. \end{cases}$$

If v happens to be smooth, then

$$\nabla\phi(x_0) = \nabla v(x_0), \quad \mathbb{D}^2\phi(x_0) \leq \mathbb{D}^2 v(x_0)$$

by the infinitesimal calculus. Here

$$\mathbb{D}^2\phi(x_0) = \left(\frac{\partial^2 \phi(x_0)}{\partial x_i \partial x_j} \right)_{n \times n}$$

is the Hessian matrix evaluated at the touching point x_0. For symmetric real matrices we use the ordering

---

[1] A few chapters from [K] are enough for our purpose.

© The Author(s), under exclusive license to Springer Nature Switzerland AG 2019
P. Lindqvist, *Notes on the Stationary p-Laplace Equation*,
SpringerBriefs in Mathematics, https://doi.org/10.1007/978-3-030-14501-9_9

$$\mathbb{X} \leq \mathbb{Y} \iff \sum_{i,j}^{n} x_{ij}\xi_i\xi_j \leq \sum_{i,j}^{n} y_{ij}\xi_i\xi_j \quad \text{for all} \quad \xi = (\xi_1, \xi_2, \ldots, \xi_n).$$

In particular, it follows that

$$\nabla v(x_0) = \nabla \phi(x_0), \quad \frac{\partial^2 v(x_0)}{\partial x_i^2} \geq \frac{\partial^2 \phi(x_0)}{\partial x_i^2}, \quad i = 1, 2, \ldots, n,$$

and so

$$\Delta v(x_0) \geq \Delta \phi(x_0), \quad \Delta_\infty v(x_0) \geq \Delta_\infty \phi(x_0), \quad \Delta_p v(x_0) \geq \Delta_p \phi(x_0).$$

If v is a p-superharmonic function and $v \in C^2(\Omega)$, then $\Delta_p v \leq 0$ and hence

$$\Delta_p \phi(x_0) \leq 0, \quad \text{if} \quad p > 2.$$

Notice carefully that the last inequality makes sense, even if v does not have any derivatives! The function ϕ has. The next definition is based on this important observation.

Definition 9.1 Let $p \geq 2$. We say that $v : \Omega \to (-\infty, \infty]$ is a *viscosity supersolution* of the p-Laplace Equation, if

  (i) v is finite in a dense subset
 (ii) v is lower semicontinuous
(iii)   whenever $\phi \in C^2(\Omega)$ touches v from below at the point $x_0 \in \Omega$, we have

$$\Delta_p \phi(x_0) \leq 0.$$

Definition 9.2 Let $p \geq 2$. We say that $u : \Omega \to [-\infty, \infty)$ is a *viscosity subsolution* of the p-Laplace Equation, if

  (i) u is finite in a dense subset
 (ii) u is upper semicontinuous
(iii)   whenever $\psi \in C^2(\Omega)$ touches u from above at the point $x_0 \in \Omega$, we have

$$\Delta_p \psi(x_0) \geq 0.$$

Definition 9.3 We say that $h \in C(\Omega)$ is a *viscosity solution* in Ω if it is both a viscosity subsolution and a viscosity supersolution.

Remark

- The operator $\Delta_p \phi$ or $\Delta_p \psi$ is evaluated only at the point x_0 of contact (the touching point). Each point has its own family of test functions. If there is no test function touching at x_0, then there is no requirement to be verified; the point passes for

free. (It is not difficult to prove that the possible touching points are dense in
Ω.) *Aide-mémoire*: supersolutions are touched from *below* since they are *lower*
semicontinuous.

- The *strict* touching $\phi(x) < v(x)$, $x \neq x_0$, can be replaced by $\phi(x) \leq v(x)$.
- The viscosity solution[2] is *assumed* to be continuous, a property that required a
  proof for weak solutions!

If $1 < p < 2$, the operator

$$\Delta_p \phi = |\nabla \phi|^{p-2} \Delta \phi + (p-2)|\nabla \phi|^{p-4} \Delta_\infty \phi$$

is undefined at the critical points, the points where $\nabla \phi = 0$. For this equation the
theory works, if one adds the requirement that $\nabla \phi(x_0) \neq 0$ at the point of contact,
cf. [JLM]. Thus there is no condition to be verified at the critical points. For $p \geq 2$
this amendment is of no influence, since $\Delta_p \phi(x_0) = 0$ now.

The definition is consistent. If $v \in C^2(\Omega)$ is a viscosity supersolution then it is
also a weak supersolution. This is clear if $p \geq 2$ since v itself then can serve as a
test function.

Theorem 9.4 *A p-superharmonic function is a viscosity supersolution.*

Proof Let v be a p-superharmonic function in the domain Ω. In order to prove that
it satisfies Definition 9.1 we use an indirect reasoning. Assume that there is a test
function ϕ touching v from below at the point x_0 and satisfying $\Delta_p \phi(x_0) > 0$. By
continuity

$$\Delta_p \phi(x) > 0 \quad \text{in} \quad B(x_0, \delta)$$

for some small δ. We may also assume that the touching is strict: $v(x) > \phi(x)$ when
$x \neq x_0$. Now ϕ is a p-subharmonic function in $B(x_0, \delta)$ and by adding the constant

$$m = \frac{1}{2} \min_{\partial B(x_0, \delta)} (v - \phi) > 0$$

we obtain the p-subharmonic function $\phi(x) + m$ satisfying

$$\phi(x) + m \leq v(x) \quad \text{on} \quad \partial B(x_0, \delta).$$

By the Comparison Principle $\phi(x) + m \leq v(x)$ in $B(x_0, \delta)$, but this is a contradiction
at the point $x = x_0$.   □

Jets. We shall need an equivalent formulation for viscosity solutions, using so-called
jets. We say that a pair $(\xi, \mathbb{X})$, where ξ is a vector in $\mathbf{R}^n$ and $\mathbb{X}$ is a symmetric
$n \times n$-matrix, belongs to the *subjet* $J^{2,-} v(x)$ if

---

[2]Here the word "viscosity" is only a label. It comes from the method of vanishing viscosity. In our
case one would replace $\Delta_p u = 0$ by $\Delta_p u_\varepsilon + \varepsilon \Delta u_\varepsilon = 0$ and send ε to zero, so that the artificial
viscosity term $\varepsilon \Delta u_\varepsilon$ vanishes. So $\lim u_\varepsilon = u$ is reached. Properly arranged, this is the same concept.

$$v(y) \geq v(x) + \langle \xi, y - x \rangle + \frac{1}{2} \langle y - x, \mathbb{X}(y - x) \rangle + o(|y - x|^2)$$

as $y \to x$. See [K], Sect. 2.2, p. 17. If it so happens that v has continuous second derivatives, then we can take $\xi = \nabla v(x)$, $\mathbb{X} = \mathbb{D}^2 v(x) =$ the Hessian matrix. In other words,

$$(\nabla v(x), \mathbb{D}^2 v(x)) \in J^{2,-} v(x), \quad \text{if} \quad v \in C^2(\Omega),$$

and the polynomial in y is Taylor's. As we shall see later, it is important that the second Alexandrov derivatives will do as members of the subjet. We need a necessary[3] condition for subjets.

Lemma 9.5 *Let* $p \geq 2$. *If* $\Delta_p v \leq 0$ *in the viscosity sense, then*

$$|\xi|^{p-2} \mathsf{Trace}(\mathbb{X}) + (p - 2)|\xi|^{p-4} \langle \xi, \mathbb{X} \xi \rangle \ \leq \ 0$$

when $(\xi, \mathbb{X}) \in J^{2,-} v(x)$ *and* $x \in \Omega$.

Infimal Convolutions. Assume that the function v is lower semicontinuous and that

$$0 \leq v(x) \leq L.$$

Define again the *infimal convolution*

$$v_\varepsilon(x) = \inf_{y \in \Omega} \left\{ v(y) + \frac{|y - x|^2}{2\varepsilon} \right\}.$$

and recall its properties in Sect. 5.3. We prove Lemma 5.12 again, but this time for viscosity supersolutions.

Theorem 9.6 *Assume that* $0 \leq v \leq L$ *in* Ω. *If* v *is a viscosity supersolution in* Ω, *so is the infimal convolution* v_ε *in the open set*

$$\Omega_\varepsilon = \left\{ x \in \Omega | \operatorname{dist}(x, \partial\Omega) > \sqrt{2L\varepsilon} \right\}.$$

Proof If x belongs to Ω_ε then the infimum in the definition of v_ε is attained at some point $x^* \in \Omega$. That x^* cannot escape to the boundary was shown in the proof of Lemma 5.12.

Fix $x_0 \in \Omega_\varepsilon$. Assume that the test function ϕ touches v_ε from below at x_0. Then x_0^* is in Ω. Using

$$\phi(x_0) = v_\varepsilon(x_0) = \frac{|x - x_0|^2}{2\varepsilon} + v(x_0^*)$$

$$\phi(x) \leq v_\varepsilon(x) \leq \frac{|x - y|^2}{2\varepsilon} + v(y)$$

---

[3]It is also sufficient provided that the closures $\overline{J^{2,-}} v$ of the subjets are evoked. See [K].

we can verify that the test function

$$\psi(x) = \phi(x + x_0 - x_0^*) - \frac{|x - x_0^*|^2}{2\varepsilon}$$

touches the original function v from below at x_0^*. At this interior point $\Delta_p \psi(x_0^*) \leq 0$ by assumption. Because

$$\nabla \psi(x_0^*) = \nabla \phi(x_0), \quad \mathbb{D}^2 \psi(x_0^*) = \mathbb{D}^2 \psi(x_0)$$

we also have

$$\Delta_p \phi(x_0) = \Delta_p \psi(x_0^*) \leq 0.$$

Thus v fulfills the requirement in Definition 9.1. □

We aim at proving that the viscosity supersolution v is a p-superharmonic function. The key step is to achieve the following result in the shrunken domain Ω_ε for the infimal convolution v_ε.

Proposition 9.7 *If v is a viscosity supersolution and $0 \leq v(x) \leq L$ in Ω, then v_ε is a weak supersolution in Ω_ε, i.e.*

$$\int_{\Omega_\varepsilon} |\nabla v_\varepsilon|^{p-2} \langle \nabla v_\varepsilon, \nabla \eta \rangle dx \geq 0 \tag{9.8}$$

when $\eta \in C_0^\infty(\Omega_\varepsilon)$, $\eta \geq 0$.

The proof in [JJ] relies on the fact that the infimal convolutions possess second derivatives in the sense of Alexandrov, which can be used in the testing with subjets. We turn to the preparations for the proof, which we shall provide only for $p \geq 2$.

Theorem 9.9 (Alexandrov) *A concave function $f : \mathbf{R}^n \to \mathbf{R}$ has second derivatives in the sense of Alexandrov: at a.e. point x there is a symmetric $n \times n$-matrix $\mathbb{A} = \mathbb{A}(x)$ such that the expansion*

$$f(y) = f(x) + \langle \nabla f(x), y - x \rangle + \frac{1}{2} \langle y - x, \mathbb{A}(x)(y - x) \rangle + o(|y - x|^2)$$

is valid as $y \to x$.

Proof We refer to [EG], Sect. 6.4, pp. 242–245. Some details in [GZ], Lemma 7.11, p. 199, are helpful to understand the singular part in the Lebesgue decomposition. □

The first derivatives are Sobolev derivatives and $\nabla f \in L_{loc}^\infty$. The problem is with the second ones. We use the notation $\mathbb{D}^2 f = \mathbb{A}$, although the Alexandrov derivatives may contain a singular Radon measure (for example, Dirac's measure) so that integration by parts can fail. The proof in [EG] establishes that a.e. we have

$$A = \lim_{\varepsilon \to 0}\left(\mathbb{D}^2(f \star \rho_\varepsilon)\right)$$

where ρ_ε is Friedrich's mollifier. This is an important ingredient in the proof. Alexandrov's Theorem is applicable to the concave function

$$f_\varepsilon(x) = v_\varepsilon(x) - \frac{|x|^2}{2\varepsilon}, \tag{9.10}$$

which is defined in the whole space (although the infimum is only over a subset). The quadratic part has no influence. Thus

$$\mathbb{D}^2 v_\varepsilon(x) = \lim_{\sigma \to 0}\left(\mathbb{D}^2(v_\varepsilon \star \rho_\sigma)(x)\right) \tag{9.11}$$

a.e. in $\mathbf{R}^n$.

We learned that the Alexandrov derivatives $\mathbb{D}^2 v_\varepsilon(x)$ exist at a.e. x. It follows that

$$\left(\nabla v_\varepsilon(x), \mathbb{D}^2 v_\varepsilon(x)\right) \in J^{2,-} v_\varepsilon(x)$$

almost everywhere. By Lemma 9.5 the inequality

$$\begin{aligned}
\Delta_p v_\varepsilon(x) &= |\nabla v_\varepsilon(x)|^{p-2}\Delta v_\varepsilon(x) \\
&\quad + (p-2)|\nabla v_\varepsilon(x)|^{p-4}\Delta_\infty v_\varepsilon(x) \\
&\leq 0
\end{aligned}$$

is valid a. e. in Ω_ε. Here $\Delta v_\varepsilon = \mathsf{Trace}(\mathbb{D}^2 v_\varepsilon)$.

We need a further mollification of the function f_ε in (9.10). Define the convolution

$$f_{\varepsilon,j} = f_\varepsilon \star \rho_{\varepsilon_j}, \qquad \rho_{\varepsilon_j}(x) = \begin{cases} \dfrac{C_n}{\varepsilon_j^n}\exp\!\left(\dfrac{-\varepsilon_j^2}{\varepsilon_j^2 - |x|^2}\right), & |x| < \varepsilon_j \\[2mm] 0, & \text{otherwise.} \end{cases}$$

The smooth functions $v_{\varepsilon,j} = v_\varepsilon \star \rho_{\varepsilon_j}$ satisfy

$$\int_{\Omega_\varepsilon} \langle |\nabla v_{\varepsilon,j}|^{p-2}\nabla v_{\varepsilon,j}, \nabla\eta \rangle \, dx = \int_{\Omega_\varepsilon} \eta\left(-\Delta_p v_{\varepsilon,j}\right) dx$$

when $\eta \in C_0^\infty(\Omega_\varepsilon)$. Keep $\eta \geq 0$. We want to extend this to v_ε. Notice that the functions $v_{\varepsilon,j}$ are *not* viscosity supersolutions themselves. By (9.11)

$$\lim_{j \to \infty} \mathbb{D}^2 v_{\varepsilon,j}(x) = \mathbb{D}^2 v_\varepsilon(x)$$

almost everywhere. Thus also

$$\lim_{j \to \infty} \Delta_p v_{\varepsilon,j}(x) = \Delta_p v_\varepsilon(x)$$

at almost every x in Ω_ε. The convolution has preserved the concavity so that $\mathbb{D}^2 f_{\varepsilon,j} \leq 0$. Since $\mathbb{D}^2 |x|^2 = +2I_n$, we obtain

$$\mathbb{D}^2 v_{\varepsilon,j} \leq \frac{I_n}{\varepsilon}, \qquad \Delta v_{\varepsilon,j} \leq \frac{n}{\varepsilon},$$

where $I_n = (\delta_{ij})$ is the unit matrix. It is immediate that

$$|\nabla v_{\varepsilon,j}| \leq \|\nabla v_\varepsilon\|_{\infty,\Omega_\varepsilon} = C_\varepsilon$$

in the support of η.

Together these inequalities yield the bound

$$-\Delta_p v_{\varepsilon,j} \geq -C_\varepsilon^{p-2} \frac{n+p-2}{\varepsilon}$$

valid almost everywhere in the support of η. This lower bound justifies the use of Fatou's Lemma below:

$$\begin{aligned}
\int_{\Omega_\varepsilon} \langle |\nabla v_\varepsilon|^{p-2} \nabla v_\varepsilon, \nabla \eta \rangle \, dx &= \lim_{j \to \infty} \int_{\Omega_\varepsilon} \langle |\nabla v_{\varepsilon,j}|^{p-2} \nabla v_{\varepsilon,j}, \nabla \eta \rangle \, dx \\
&= \lim_{j \to \infty} \int_{\Omega_\varepsilon} \eta \left(-\Delta_p v_{\varepsilon,j} \right) dx \\
&\geq \int_{\Omega_\varepsilon} \liminf_{j \to \infty} \left(\eta \left(-\Delta_p v_{\varepsilon,j} \right) \right) dx \\
&= \int_{\Omega_\varepsilon} \eta \left(-\Delta_p v_\varepsilon \right) dx \\
&\geq \int_{\Omega_\varepsilon} \eta \, 0 \, dx = 0.
\end{aligned}$$

In the very last step we used the pointwise inequality $-\Delta_p v_\varepsilon \geq 0$, which we recall that needed Alexandrov's Theorem for its proof.

Now the proof of Proposition 9.7 is accomplished. It remains to pass to the limit as $\varepsilon \to 0$.

Theorem 9.12 *Suppose that v is a bounded viscosity supersolution of $\Delta_p v \leq 0$ in Ω. Then the Sobolev gradient ∇v exists and $v \in W_{loc}^{1,p}(\Omega)$. Furthermore,*

$$\int_\Omega \langle |\nabla v|^{p-2} \nabla v, \nabla \eta \rangle \, dx \geq 0$$

for all $\eta \geq 0$, $\eta \in C_0^\infty(\Omega)$.

Proof Choose $\zeta \geq 0$, $\zeta \in C_0^\infty(\Omega)$. We can assume $0 \leq v(x) \leq L$ in the support of ζ. Use

$$\eta(x) = (L - v_\varepsilon(x))\zeta(x)^p$$

in Proposition 9.7. When ε is small enough for the inclusion supp(η) $\subset \Omega_\varepsilon$ we proceed as in the proof of Theorem 5.8. The Caccioppoli estimate for v_ε now reads

$$\int_\Omega \zeta^p |\nabla v_\varepsilon|^p \, dx \;\leq\; (pL)^p \int_\Omega |\nabla \zeta|^p \, dx.$$

By a compactness argument (at least for a subsequence) $\nabla v_\varepsilon \rightharpoonup \mathbf{w}$ weakly in $L_{loc}^p(\Omega)$. Since $v_\varepsilon \nearrow v$ we can identify $\mathbf{w} = \nabla v$. The conclusion is that ∇v exists and by weak lower semicontinuity of convex integrals also

$$\int_\Omega \zeta^p |\nabla v|^p \, dx \;\leq\; (pL)^p \int_\Omega |\nabla \zeta|^p \, dx.$$

To conclude the proof we show that the convergence $\nabla v_\varepsilon \to \nabla v$ is strong in $L_{loc}^p(\Omega)$ so that one may pass to the limit under the integral sign in (9.8). The procedure is almost a repetition of the proof of Theorem 5.8. To this end, fix a function $\theta \in C_0^\infty(\Omega)$, $0 \leq \theta \leq 1$, and use the test function $\eta = (v - v_\varepsilon)\theta$ in the equation for v_ε. Then

$$\int_\Omega \langle |\nabla v|^{p-2}\nabla v - |\nabla v_\varepsilon|^{p-2}\nabla v_\varepsilon, \nabla((v - v_\varepsilon)\theta)\rangle \, dx$$

$$\leq \int_\Omega \langle |\nabla v|^{p-2}\nabla v, \nabla((v - v_\varepsilon)\theta)\rangle \, dx \quad \longrightarrow \quad 0,$$

where the last integral approaches zero due to the weak convergence.

We split the first integral into two parts

$$\int_\Omega \theta \, \langle |\nabla v|^{p-2}\nabla v - |\nabla v_\varepsilon|^{p-2}\nabla v_\varepsilon, \nabla v - \nabla v_\varepsilon\rangle \, dx$$

$$+ \int_\Omega (v - v_\varepsilon)\langle |\nabla v|^{p-2}\nabla v - |\nabla v_\varepsilon|^{p-2}\nabla v_\varepsilon, \nabla\theta\rangle \, dx$$

and notice that the second integral approaches zero, because its absolute value is less than

$$\|v - v_\varepsilon\|_{L^p(D)} \left(\|\nabla v\|_{L^p(D)}^{p-1} + \|\nabla v_\varepsilon\|_{L^p(D)}^{p-1} \right) \|\nabla\theta\|,$$

where D contains the support of θ and $\|v - v_\varepsilon\|_{L^p(D)} \to 0$. Recall also that $\|\nabla v_\varepsilon\|_{L^p(D)}$ is uniformly bounded.

Thus we have established that

$$\lim_{\varepsilon \to 0} \int_\Omega \theta \, \langle |\nabla v|^{p-2}\nabla v - |\nabla v_\varepsilon|^{p-2}\nabla v_\varepsilon, \, \nabla v - \nabla v_\varepsilon \rangle \, dx \quad \longrightarrow \quad 0$$

at least for a subsequence. Now the strong convergence of the gradients follows. Indeed, the case $p \geq 2$ follows from inequality (I) in Chap. 12. The singular case $p < 2$ requires some work.[4] Therefore we can proceed to the limit under the integral sign in (9.8). □

Unbounded Viscosity Solutions. Let $v(x) \geq 0$ be a viscosity solution. So is the function

$$v_L(x) = \min\{v(x), L\}$$

cut at height L. The point is that v_L is *bounded*. Now Theorem 9.12 is applicable and we conclude that v_L is a p-superharmonic function. So is v itself as a limit of an increasing sequence: $v = \lim v_L$. Indeed, to verify the Comparison Principle for v, we take $D \subset\subset \Omega$ and assume that $h \in C(\overline{D})$ is a p-harmonic function with boundary values $h|_{\partial D} \leq v|_{\partial D}$. Take $L > \max\{h\}$. Then $v_L \geq h$ on ∂D. By the Comparison Principle, which is valid for v_L, one has $v_L \geq h$ in D. Since $v \geq v_L$, we conclude that $v \geq h$ in D. This proves the Comparison Principle for v.

We assumed that $v \geq 0$ but since the theorem is local, this is no restriction. We have now proven the converse of Theorem 9.4.

Theorem 9.13 *A viscosity supersolution of $\Delta_p v \leq 0$ is a p-superharmonic function.*

The Theorem has an interesting consequence. A viscosity supersolution v has a gradient ∇v in Sobolev's sense and $\nabla v \in L^s_{loc}$ for some exponent s described in Theorem 5.18. This gradient was not mentioned in Definition 9.1!

4

$$\int |\nabla v - \nabla v_\varepsilon|^p \, dx$$

$$= \int |\nabla v - \nabla v_\varepsilon|^p \left(1 + |\nabla v|^2 + |\nabla v_\varepsilon|^2\right)^{p(p-2)/4} \left(1 + |\nabla v|^2 + |\nabla v_\varepsilon|^2\right)^{p(2-p)/4} \, dx$$

$$\leq \left\{ \int |\nabla v - \nabla v_\varepsilon|^2 \left(1 + |\nabla v|^2 + |\nabla v_\varepsilon|^2\right)^{(p-2)/2} \right\}^{\frac{p}{2}} \left\{ \int \left(1 + |\nabla v|^2 + |\nabla v_\varepsilon|^2\right)^{p/2} \, dx \right\}^{1-\frac{p}{2}}$$

$$\leq \left\{ \frac{1}{p-1} \int \langle |\nabla v|^{p-2}\nabla v - |\nabla v_\varepsilon|^{p-2}\nabla v_\varepsilon, \, \nabla v - \nabla v_\varepsilon \rangle \, dx \right\}^{\frac{p}{2}} \left\{ \int \left(1 + |\nabla v|^2 + |\nabla v_\varepsilon|^2\right)^{p/2} \, dx \right\}^{\frac{2-p}{2}}$$

where inequality (VII) in Sect. 12 was used at the last step.

Chapter 10
Asymptotic Mean Values

The celebrated Mean Value Property, discovered by Gauss for harmonic functions, has a sophisticated replacement for p-harmonic functions. Naturally, a non linear extra term is needed. Furthermore, the new formula is valid in a particular asymptotic (infinitesimal) sense. The mean values

$$\fint_{B(x,\varepsilon)} u(y)\,dy \equiv \frac{1}{|B(x,\varepsilon)|} \int_{B(x,\varepsilon)} u(y)\,dy$$

are taken over balls with radii ε shrinking to 0. W. Blaschke observed in 1916 that a continuous function u is harmonic in Ω if and only if

$$u(x) = \fint_{B(x,\varepsilon)} u(y)\,dy + o(\varepsilon^2) \quad \text{as} \quad \varepsilon \to 0$$

when $x \in \Omega$. Here $\varepsilon^{-2} o(\varepsilon^2) \to 0$. (We deliberately ignore the fact that here the error term $o(\varepsilon^2) \equiv 0$ a posteriori.)

The Fundamental Asymptotic Formula

$$u(x) = \frac{p-2}{p+n} \frac{\sup_{B(x,\varepsilon)} u + \inf_{B(x,\varepsilon)} u}{2} + \frac{2+n}{p+n} \fint_{B(x,\varepsilon)} u(y)\,dy + o(\varepsilon^2) \tag{10.1}$$

was given by Manfredi, Parviainen, and Rossi in [MPR1].[1] As we shall see, properly interpreted in the viscosity sense, it characterizes the p-harmonic functions. The first term counts for the nonlinearity and the second one (the mean value) is linear. When $p = \infty$ one should read

---

[1] This Section is based on [MPR1].

P. Lindqvist, *Notes on the Stationary p-Laplace Equation*,
SpringerBriefs in Mathematics, https://doi.org/10.1007/978-3-030-14501-9_10

$$u(x) = \frac{1}{2}\left(\sup_{B(x,\varepsilon)} u + \inf_{B(x,\varepsilon)} u\right) + o(\varepsilon^2). \tag{10.2}$$

For an interpretation in Stochastic Game Theory it is essential that the coefficients sum up to 1, i.e.

$$\frac{p-2}{p+n} + \frac{2+n}{p+n} = 1.$$

They represent probabilities, at least for $p > 2$.

Smooth Functions. Let us first derive some formulas for sufficiently smooth functions, say $\phi \in C^2(\Omega)$. Integrating the Taylor formula

$$\phi(y) = \phi(x) + \nabla\phi(x) \cdot (y - x) + \frac{1}{2}\sum_{i,j=1}^{n}\frac{\partial^2\phi(x)}{\partial x_i \partial x_j}(y_i - x_i)(y_j - x_j)$$

$$+ o(|x - y|^2)$$

with respect to y over $B(x, \varepsilon)$ we get

$$\fint_{B(x,\varepsilon)}\phi(y)\,dy = \phi(x) + 0 + \frac{1}{2}\sum_{i,j=1}^{n}\frac{\partial^2\phi(x)}{\partial x_i \partial x_j}\fint_{B(x,\varepsilon)}(y_i - x_i)(y_j - x_j)\,dy + o(\varepsilon^2).$$

By symmetric cancellation the integrals vanish for $i \neq j$, and for $i = j$ we have

$$\fint_{B(x,\varepsilon)}(y_i - x_i)^2 dy = \frac{1}{n}\fint_{B(x,\varepsilon)}|y - x|^2 dy = \frac{\varepsilon^2}{n+2}.$$

Thus we have arrived at the asymptotic formula in the lemma below. (It is a truncated version of the Pizzetti formula from 1909.)

Lemma 10.3 *Let $\phi \in C^2(\Omega)$. When $x \in \Omega$*

$$\phi(x) = \fint_{B(x,\varepsilon)}\phi(y)\,dy - \frac{\varepsilon^2}{n+2}\Delta\phi(x) + o(\varepsilon^2) \tag{10.4}$$

as $\varepsilon \to 0$.

In the next lemma the critical points must be avoided. There the *normalized* ∞-Laplacian appears:

$$\Delta_\infty^\diamond \phi \equiv |\nabla\phi|^{-2}\Delta_\infty\phi = \left\langle \mathbb{D}^2\phi\frac{\nabla\phi}{|\nabla\phi|}, \frac{\nabla\phi}{|\nabla\phi|}\right\rangle.$$

Later the normalized p-Laplace operator

$$\Delta_p^\diamond \phi \equiv |\nabla\phi|^{2-p}\Delta_p\phi = \Delta\phi + (p - 2)\Delta_\infty^\diamond \phi$$

will be useful.[2]

Lemma 10.5 *Let $\phi \in C^2(\Omega)$. If $x \in \Omega$ and $\nabla\phi(x) \neq 0$, then*

$$\phi(x) = \frac{1}{2}\left(\max_{\overline{B}(x,\varepsilon)} \phi + \min_{\overline{B}(x,\varepsilon)} \phi\right) - \frac{\Delta_\infty\phi(x)}{2\,|\nabla\phi(x)|^2}\,\varepsilon^2 + o(\varepsilon^2) \tag{10.6}$$

as $\varepsilon \to 0$.

Proof Let $x_0 \in \Omega$ with $\nabla\phi(x_0) \neq 0$. Choose $\varepsilon > 0$ so small that

$$\nabla\phi(y) \neq 0 \quad\text{when}\quad |y - x_0| \leq \varepsilon.$$

The extremal values are attained at points x on $\partial B(x_0, \varepsilon)$ where

$$x = x_0 \pm \varepsilon\,\frac{\nabla\phi(x)}{|\nabla\phi(x)|}, \qquad |x - x_0| = \varepsilon. \tag{10.7}$$

The plus sign corresponds to the maximum and the minus sign to the minimum. To see this, use Lagrange multipliers. They are *approximatively* opposite endpoints of a diameter.[3] It follows from

$$\phi_{x_j}(y) = \phi_{x_j}(x_0) + O(\varepsilon)$$

that

$$\frac{\nabla\phi(y)}{|\nabla\phi(y)|} = \frac{\nabla\phi(x_0)}{|\nabla\phi(x_0)|} + O(\varepsilon), \quad\text{when}\quad |y - x_0| \leq \varepsilon.$$

Let $x, x^* \in \partial B(x_0, \varepsilon)$ be *exactly* the endpoints of a diameter, i.e.

$$\frac{x + x^*}{2} = x_0.$$

Add the two Taylor expansions

$$\phi(y) = \phi(x_0) + \langle\nabla\phi(x_0), y - x_0\rangle + \frac{1}{2}\langle\mathbb{D}^2\phi(x_0)(y - x_0), y - x_0\rangle$$
$$+ o(|y - x_0|^2)$$

for $y = x$ and $y = x^*$. The first order gradient terms cancel so that

$$\phi(x) + \phi(x_0) = 2\phi(x_0) + \langle\mathbb{D}^2\phi(x_0)(y - x_0), y - x_0\rangle + o(|y - x_0|^2).$$

---

[2]Clarification: In some recent literature, careless use of the traditional symbol Δ_p also for the normalized operator creates confusion about the proper meaning! Often Δ_p^N or Δ_p^G are used to distinguish the normalized operator.

[3]The notation hides the dependence on ε, i.e. $x = x^\varepsilon$.

In the quadratic term we substitute

$$x - x_0 = \pm\varepsilon \frac{\nabla\phi(x)}{|\nabla\phi(x)|} = \pm\varepsilon \frac{\nabla\phi(x_0)}{|\nabla\phi(x_0)|} + O(\varepsilon^2).$$

It follows that

$$\phi(x) + \phi(x_0) = 2\phi(x_0) + \varepsilon^2 \Delta_\infty^\diamond \phi(x_0) + o(\varepsilon^2),$$

where $x + x^* = 2x_0$ and x satisfies (10.7).

First, let x be the maximum point:

$$\phi(x) = \max_{\overline{B}(x_0,\varepsilon)} \phi.$$

Then

$$\max_{\overline{B}(x_0,\varepsilon)} \phi + \min_{\overline{B}(x_0,\varepsilon)} \phi = \phi(x) + \min_{\overline{B}(x_0,\varepsilon)} \phi$$
$$\leq \phi(x) + \phi(x^*)$$
$$\leq 2\phi(x_0) + \varepsilon^2 \Delta_\infty^\diamond \phi(x_0) + o(\varepsilon^2).$$

We can derive the opposite inequality by selecting x as the minimum point, since now

$$\max_{\overline{B}(x_0,\varepsilon)} \phi + \min_{\overline{B}(x_0,\varepsilon)} \phi \geq \phi(x^*) + \phi(x).$$

This proves formula (10.6) with x_0 in the place of x. $\square$

Lemma 10.8 *Let $\phi \in C^2(\Omega)$. If $\nabla\phi(x) \neq 0$ at the point $x \in \Omega$, then the asymptotic expansion*

$$\phi(x) = \frac{p-2}{p+n} \frac{\max_{\overline{B}(x,\varepsilon)} \phi + \min_{\overline{B}(x,\varepsilon)} \phi}{2} + \frac{2+n}{p+n} \fint_{B(x,\varepsilon)} \phi(y)\,dy$$
$$- \frac{1}{2}\varepsilon^2 \Delta_p^\diamond \phi(x) + o(\varepsilon^2) \tag{10.9}$$

is valid as $\varepsilon \to 0$.

Proof Combine the asymptotic formulas (10.4) and (10.6) in a suitable way to see this. $\square$

Mean values in the viscosity sense. We can read off from formula (10.9) that a function $u \in C^2(\Omega)$ such that $\nabla u \neq 0$ in Ω is p-harmonic in Ω if and only if the fundamental asymptotic formula (10.1) holds in Ω. Unfortunately, in the presence of critical points (i.e. zeros of the gradient) the above calculation is not valid. Recall that $\Delta_p u = 0$ in the viscosity sense if and only if

$$|\nabla u|^2 \Delta u + (p-2)\Delta_\infty u = 0$$

in the viscosity sense. No testing is needed at the critical points of touching test functions. On the other hand, we know that the p-harmonic functions are exactly the viscosity solutions of the p-Laplace Equation. This suggests to interpret also the fundamental asymptotic formula (10.1) in the viscosity sense.

Definition 10.10   A function u satisfies the formula

$$u(x) = \frac{p-2}{p+n}\frac{\max_{\overline{B}(x,\varepsilon)} u + \min_{\overline{B}(x,\varepsilon)} u}{2} + \frac{2+n}{p+n}\fint_{B(x,\varepsilon)} u(y)\,dy + o(\varepsilon^2)$$

in the viscosity sense, if the two conditions:

- If $x_0 \in \Omega$ and if $\phi \in C^2(\Omega)$ touches u from below at x_0, then

$$\phi(x) \geq \frac{p-2}{p+n}\frac{\max_{\overline{B}(x,\varepsilon)} \phi + \min_{\overline{B}(x,\varepsilon)} \phi}{2} + \frac{2+n}{p+n}\fint_{B(x,\varepsilon)} \phi(y)\,dy + o(\varepsilon^2)$$

  as $\varepsilon \to 0$. Furthermore, if it so happens that $\nabla\phi(x_0) = 0$ then the test function must obey the rule $\mathbb{D}^2\phi(x_0) \leq 0$.

- If $x_0 \in \Omega$ and if $\psi \in C^2(\Omega)$ touches u from below at x_0, then

$$\psi(x) \leq \frac{p-2}{p+n}\frac{\max_{\overline{B}(x,\varepsilon)} \psi + \min_{\overline{B}(x,\varepsilon)} \psi}{2} + \frac{2+n}{p+n}\fint_{B(x,\varepsilon)} \psi(y)\,dy + o(\varepsilon^2)$$

  as $\varepsilon \to 0$. Furthermore, if it so happens that $\nabla\psi(x_0) = 0$ then the test function must obey the rule $\mathbb{D}^2\psi(x_0) \geq 0$.

hold.

Remark The extra restrictions on the test functions at critical points are used to conclude that

$$\lim_{y\to x_0}\frac{\phi(y)-\phi(x_0)}{|y-x_0|^2} \leq 0, \qquad \lim_{y\to x_0}\frac{\psi(y)-\psi(x_0)}{|y-x_0|^2} \geq 0.$$

Theorem 10.11 (Manfredi - Parviainen -Rossi) *Let $u \in C(\Omega)$. Then $\Delta_p u = 0$ in the viscosity sense if and only if the Asymptotic Mean Value Formula (10.1) holds in the viscosity sense.*

Proof Let us consider the case of subsolutions. Thus we assume that $\psi \in C^2(\Omega)$ touches u from above at the point $x_0 \in \Omega$. If $\nabla\psi(x_0) \neq 0$, then the desired conclusions are contained in Lemma 10.8.

In the situation $\nabla\psi(x_0) = 0$ we proceed as follows. First, we assume that $\Delta_p u \geq 0$ in the viscosity sense. We have to verify that

$$\psi(x_0) \leq \frac{p-2}{p+n} \frac{\max\limits_{\overline{B}(x_0,\varepsilon)} \psi + \min\limits_{\overline{B}(x_0,\varepsilon)} \psi}{2} + \frac{2+n}{p+n} \fint_{B(x_0,\varepsilon)} \psi(y)\, dy + o(\varepsilon^2). \quad (10.12)$$

Now only the situation $\mathbb{D}^2\psi(x_0) \geq 0$ is permissible. In particular, $\Delta\psi(x_0) \geq 0$ and hence

$$\psi(x_0) \leq \fint_{B(x_0,\varepsilon)} \psi(y)\, dy + o(\varepsilon^2) \quad (10.13)$$

by (10.4). The extra condition

$$\lim_{y \to x_0} \frac{\psi(y) - \psi(x_0)}{|y - x_0|^2} \geq 0$$

is at our disposal. Denoting

$$\psi(x_\varepsilon) = \min_{|y - x_0| \leq \varepsilon} \psi(y)$$

we have

$$\liminf_{\varepsilon \to 0} \frac{1}{\varepsilon^2} \left\{ \frac{1}{2}\left(\max_{\overline{B}(x_0,\varepsilon)} \psi + \min_{\overline{B}(x_0,\varepsilon)} \psi\right) - \psi(x_0) \right\}$$

$$= \liminf_{\varepsilon \to 0} \frac{1}{\varepsilon^2} \left\{ \frac{1}{2}\left(\max_{\overline{B}(x_0,\varepsilon)} \psi - \psi(x_0)\right) + \frac{1}{2}\left(\min_{\overline{B}(x_0,\varepsilon)} \psi - \psi(x_0)\right) \right\}$$

$$\geq \liminf_{\varepsilon \to 0} \frac{1}{\varepsilon^2} \left\{ \frac{1}{2}\left(\max_{\overline{B}(x_0,\varepsilon)} \psi - \psi(x_0)\right) \right\}$$

$$= \frac{1}{2} \liminf_{\varepsilon \to 0} \left(\frac{\psi(x_\varepsilon) - \psi(x_0)}{|x_\varepsilon - x_0|^2} \right)\left(\frac{|x_\varepsilon - x_0|^2}{\varepsilon^2} \right)$$

$$\geq 0$$

since $|x_\varepsilon - x_0|^2/\varepsilon^2 \leq 1$. This shows that

$$\psi(x_0) \leq \frac{1}{2}\left(\max \psi + \min \psi\right) + o(\varepsilon^2). \quad (10.14)$$

Combining (10.13) and (10.14) we see that (10.12) is valid, as desired.

Second, assume that (10.12) holds. But the mere fact that $\nabla\psi(x_0) = 0$ means that there is nothing to prove in Definition 9.2 concerning $\Delta_p\psi(x_0) \leq 0$, since critical contact points of the test function pass for free. (Therefore the Asymptotic Mean Value Formula was not even used at this step.) This concludes the case of subsolutions.

Finally, the case when the test functions touch from below is similar.   □

Comments. In the case $p = \infty$ the *Asymptotic Mean Value Formula is not valid pointwise*. The example with the ∞-harmonic function

$$u(x, y) = x^{\frac{4}{3}} - y^{\frac{4}{3}}$$

exhibits this fact, see [MPR1] for the calculations. However, in the cases $1 < p < \infty$ the Asymptotic Mean Value Formula holds pointwise in the plane ($n = 2$), cf. [LiM2] and [AL1]. The proofs in the plane are based on the hodograph method. To the best of my knowledge the situation in higher dimensions $n \geq 2$ is an open problem.

Dynamic Programming Principle. An interpretation in Stochastic Game Theory comes from the Dynamic Programming Principle or DPP that is satisfied by the function of the game:

$$u_\varepsilon(x) = \frac{\alpha}{2}\left(\sup_{B(x,\varepsilon)} u_\varepsilon + \inf_{B(x,\varepsilon)} u_\varepsilon(x)\right) + \beta \fint_{B(x,\varepsilon)} u_\varepsilon(y)\,dy$$

where the 'probabilities' are

$$\alpha = \frac{p-2}{p+n}, \qquad \beta = \frac{2+n}{p+n}.$$

The set up and how u_ε approaches a p-harmonic function is described in [MPR2]. See also [KMP] for $p < 2$. Actually, the Stochastic Game was found first, then came the Asymptotic Mean Value Formula.

Chapter 11
Some Open Problems

As a challenge we mention some problems which, to the best of our knowledge, are open for the p-Laplace equation, when $p \neq 2$. In general, the situation in the plane is better understood than in higher dimensional spaces.

The Problem of Unique Continuation. Can two different p-harmonic functions coincide in an open subset of their common domain of definition? The most pregnant version is the following. Suppose that $u = u(x_1, x_2, x_3)$ is p-harmonic in $\mathbf{R}^3$ and that $u(x_1, x_2, x_3) = 0$ at each point in the lower half-space $x_3 < 0$. Is $u \equiv 0$ then? The plane case $n = 2$ is solved in [BI1]. In the extreme case $p = \infty$, the Principle of Unique Continuation does not hold. —See also [M].

The Strong Comparison Principle. Suppose that u_1 and u_2 are p-harmonic functions satisfying $u_2 \geq u_1$ in the domain Ω. If $u_2(x_0) = u_1(x_0)$ at some interior point x_0 of Ω, does it follow that $u_2 \equiv u_1$? The plane case is solved in [M1]. The Strong Comparison Principle does not hold for $p = \infty$. One may add that, if one of the functions is identically zero, then this is the Strong Maximum Principle, which, indeed, is valid for $1 < p \leq \infty$.

Very Weak Solutions. Suppose that $u \in W^{1,p-1}(\Omega)$ and that

$$\int_\Omega \langle |\nabla u|^{p-2} \nabla u, \nabla \varphi \rangle dx = 0$$

for all $\varphi \in C_0^\infty(\Omega)$. Does this imply that u is (equivalent to) a p-harmonic function? Please, notice that the assumption

$$\int_\Omega |\nabla u|^{p-1} dx < \infty$$

with the exponent $p - 1$ instead of the natural exponent p is not strong enough to allow test functions like $\zeta^p u$. When $p = 2$ a stronger theorem (Weyl's lemma) holds. T. Iwaniec and G. Martin have proved that the assumption $u \in W^{1,p-\varepsilon}(\Omega)$ is sufficient for some small $\varepsilon > 0$, cf [I]. J. Lewis has given a simpler proof in [Le3].

Second Derivatives. For $1 < p < 2$ the second derivatives of a p-harmonic function are locally summable, i.e. $u \in W^{2,p}_{loc}(\Omega)$. Indeed, even $u \in W^{2,2}_{loc}(\Omega)$. What is the situation for $p > 2$? The two dimensional case $n = 2$ was settled in Sect. 7. The range $1 < p < 3 + \frac{2}{n-2}$, in which the Cordes condition is valid, was settled in [MW].

The C^1-regularity for $p = \infty$. Does an ∞-harmonic function belong to C^1_{loc}? What about $C^{1,\alpha}_{loc}$? Recently, O. Savin proved that in the plane all ∞-harmonic functions have continuous gradients, cf [Sa]. An educated guess is that the optimal regularity class is $C^{1,1/3}_{loc}$ in the plane. The Hölder exponent $1/3$ for the gradient is attained for the function $x^{4/3} - y^{4/3}$. In space Evans and Smart have proved in [ES] that the ∞-harmonic functions are (totally) differentiable at every point. The proof provides no modulus of continuity.

The Asymptotic Mean Value Property. Does the Asymptotic Mean Value Property hold *pointwise* in space? It holds in the viscosity sense. In the plane it is known to be valid pointwise, cf. [ALl] and [LiM2].

High regularity as p is close to 1. In the plane case a p-harmonic function is of some differentiability class $C^{k(p)}_{loc}$ where $k(p) \to \infty$ as $p \to 1 + 0$. (However, solutions of the limit equation are, in general, not of class C^∞.). Does the regularity increase also for $n \geq 3$ when $p \to 1 + 0$?

There are many more problems. "Luck and chance favours the prepared mind."[1]

---

[1] Dans les champs de l'observation le hasard ne favorise que les esprits préparés. LOUIS PASTEUR.

Chapter 12
Inequalities for Vectors

Some special inequalities are helpful in the study of the p-Laplace operator. Expressions like

$$\langle |\nabla v|^{p-2}\nabla v - |\nabla u|^{p-2}\nabla u, \nabla v - \nabla u\rangle$$

are ubiquitous and hence inequalities for

$$\langle |b|^{p-2}b - |a|^{p-2}a, b - a\rangle$$

are needed, a and b denoting vectors in $\mathbf{R}^n$. As expected, the cases $p > 2$ and $p < 2$ are different. Let us begin with the identity

$$\langle |b|^{p-2}b - |a|^{p-2}a, b - a\rangle = \frac{|b|^{p-2} + |a|^{p-2}}{2}|b - a|^2$$
$$+ \frac{(|b|^{p-2} - |a|^{p-2})(|b|^2 - |a|^2)}{2},$$

which is easy to verify by a calculation. We can read off the following inequalities

(I)
$$\langle |b|^{p-2}b - |a|^{p-2}a, b - a\rangle \geq 2^{-1}(|b|^{p-2} + |a|^{p-2})|b - a|^2$$
$$\geq 2^{2-p}|b - a|^p,$$

if $p \geq 2$.

(II)
$$\langle |b|^{p-2}b - |a|^{p-2}a, b - a\rangle \leq \frac{1}{2}(|b|^{p-2} + |a|^{p-2})|b - a|^2,$$

if $p \leq 2$.

P. Lindqvist, *Notes on the Stationary p-Laplace Equation*,
SpringerBriefs in Mathematics, https://doi.org/10.1007/978-3-030-14501-9_12

However, the second inequality in (I) cannot be reversed for $p \leq 2$, as the first one, not even with a poorer constant than 2^{2-p}. Nevertheless, we have

(III)
$$\langle |b|^{p-2}b - |a|^{p-2}a, b - a \rangle \leq \gamma(p)|b - a|^p, \qquad p \leq 2,$$

according to [Db2].[1]
   The formula

$$|b|^{p-2}b - |a|^{p-2}a = \int_0^1 \frac{d}{dt}|a + t(b - a)|^{p-2}(a + t(b - a))dt$$

yields

(IV)
$$|b|^{p-2}b - |a|^{p-2}a = (b - a)\int_0^1 |a + t(b - a)|^{p-2}dt$$
$$+ (p - 2)\int_0^1 |a + t(b - a)|^{p-4}\langle a + t(b - a), b - a \rangle(a + t(b - a))dt$$

and consequently we have

$$\langle |b|^{p-2}b - |a|^{p-2}a, b - a \rangle = |b - a|^2 \int_0^1 |a + t(b - a)|^{p-2}dt$$
$$+ (p - 2)\int_0^1 |a + t(b - a)|^{p-4}\big(\langle a + t(b - a), b - a \rangle\big)^2 dt.$$

To proceed further, we notice that the last integral has the estimate

---

[1] By conjugation (III) follows from (I). To see this, let $1 < p < 2$ and write $q = p/(p - 1) > 2$.
By (I)
$$2^{2-q}|B - A|^q \leq \langle |B|^{q-2}B - |A|^{q-2}A, B - A \rangle.$$
Use $a = |A|^{q-2}A$, $A = |a|^{p-2}a$ and the same for B to obtain
$$2^{2-q}\Big||b|^{p-2}b - |a|^{p-2}a\Big|^q \leq \langle |b|^{p-2}b - |a|^{p-2}a, b - a \rangle \leq |b - a|\Big||b|^{p-2}b - |a|^{p-2}a\Big|.$$
It follows that
$$2^{2-q}\Big||b|^{p-2}b - |a|^{p-2}a\Big|^{q-1} \leq |b - a|.$$
Thus, since $(p - 1)(q - 1) = 1$,

$$\boxed{\Big||b|^{p-2}b - |a|^{p-2}a\Big| \leq 2^{2-p}|b - a|^{p-1}, \qquad 1 < p < 2.}$$

This directly implies (III) with $\gamma(p) = 2^{2-p}$.

$$0 \leq \int_0^1 |a + t(b-a)|^{p-4} \big(\langle a + t(b-a), b-a \rangle \big)^2 dt$$

$$\leq |b-a|^2 \int_0^1 |a + t(b-a)|^{p-2} dt \ .$$

We begin with $p \geq 2$. First we get

$$\langle |b|^{p-2}b - |a|^{p-2}a, b-a \rangle \geq |b-a|^2 \int_0^1 |a + t(b-a)|^{p-2} dt$$

and hence

$$\big| |b|^{p-2}b - |a|^{p-2}a \big| \geq |b-a| \int_0^1 |a + t(b-a)|^{p-2} dt$$

by the Cauchy-Schwarz inequality. We also have

$$\big| |b|^{p-2}b - |a|^{p-2}a \big| \leq (p-1)|b-a| \int_0^1 |a + t(b-a)|^{p-2} dt \ ,$$

where $p \geq 2$. Continuing, we obtain replacing p by $(p+2)/2$:

$$\big| |b|^{\frac{p-2}{2}}b - |a|^{\frac{p-2}{2}}a \big|^2 \leq \Big(\frac{p}{2}\Big)^2 |b-a|^2 \Big(\int_0^1 |a + t(b-a)|^{\frac{p-2}{2}} dt \Big)^2$$

$$\leq \Big(\frac{p}{2}\Big)^2 |b-a|^2 \int_0^1 |a + t(b-a)|^{p-2} dt \leq \Big(\frac{p}{2}\Big)^2 \langle |b|^{p-2}b - |a|^{p-2}a, b-a \rangle$$

We have arrived at

(V)

$$\big| |b|^{\frac{p-2}{2}}b - |a|^{\frac{p-2}{2}}a \big|^2 \leq \Big(\frac{p^2}{4}\Big) \langle |b|^{p-2}b - |a|^{p-2}a, b-a \rangle$$

if $p \geq 2$

This is one of the inequalities used by Bojarski and Iwaniec (see Chap. 4). We also have, keeping $p \geq 2$,

$$\big| |b|^{p-2}b - |a|^{p-2}a \big| \leq (p-1)|b-a| \int_0^1 |a + t(b-a)|^{p-2} dt$$

$$\leq (p-1)|b-a| \big(|b|^{\frac{p-2}{2}} + |a|^{\frac{p-2}{2}} \big) \int_0^1 |a + t(b-a)|^{\frac{p-2}{2}} dt$$

$$\leq (p-1) \big(|b|^{\frac{p-2}{2}} + |a|^{\frac{p-2}{2}} \big) \big| |b|^{\frac{p-2}{2}}b - |a|^{\frac{p-2}{2}}a \big| \ .$$

At the intermediate step $|a + t(b-a)|^{p-2}$ was factored and then

$$|a + t(b - a)|^{\frac{p-2}{2}} \leq |a|^{\frac{p-2}{2}} + |b|^{\frac{p-2}{2}}$$

was used. We have arrived at

(VI)

$$\left||b|^{p-2}b - |a|^{p-2}a\right| \leq$$
$$(p-1)\left(|b|^{\frac{p-2}{2}} + |a|^{\frac{p-2}{2}}\right)\left||b|^{\frac{p-2}{2}}b - |a|^{\frac{p-2}{2}}a\right|, \text{ if } p \geq 2$$

Also this inequality was used by Bojarski and Iwaniec in their differentiability proof.
Let us return to the formula below IV and consider now $1 < p \leq 2$. We obtain

$$\langle|b|^{p-2}b - |a|^{p-2}a, b - a\rangle \geq (p-1)|b - a|^2 \int_0^1 |a + t(b - a)|^{p-2}dt \, .$$

A simple estimation, taking into account that now $p - 2 < 0$, yields

(VII)

$$\langle|b|^{p-2}b - |a|^{p-2}a, b - a\rangle \geq (p-1)|b - a|^2(1 + |a|^2 + |b|^2)^{\frac{p-2}{2}}$$

if $1 \leq p \leq 2$.

Recall

(VIII)

$$\left||b|^{p-2}b - |a|^{p-2}a\right| \leq 2^{2-p}|b - a|^{p-1}$$

if $1 \leq p \leq 2$.

We remark that for many purposes the simple fact

$$\langle|b|^{p-2}b - |a|^{p-2}a, b - a\rangle > 0, \quad a \neq b,$$

valid for all p, is enough.
Finally we just mention that the inequality

$$|b|^p \geq |a|^p + p\langle|a|^{p-2}a, b - a\rangle, \quad p \geq 1,$$

expressing the convexity of the function $|x|^p$ can be sharpened. In the case $p \geq 2$
the inequality

$$|b|^p \geq |a|^p + p\langle|a|^{p-2}a, b - a\rangle + C(p)|b - a|^p$$

holds with a constant $C(p) > 0$. The case $1 < p < 2$ requires a modification of the
last term.

Literature

References

[A1]   G. Aronsson, Extension of functions satisfying Lipschitz conditions. Arkiv för Matematik **6**, 551–561 (1967)

[A2]   G. Aronsson, On the partial differential equation $u_x^2 u_{xx} + 2u_x u_y u_{xy} + u_y^2 u_{yy} = 0$. Arkiv för Matematik **7**, 395–435 (1968)

[A3]   G. Aronsson, On certain singular solutions of the partial differential equation $u_x^2 u_{xx} + 2u_x u_y u_{xy} + u_y^2 u_{yy} = 0$. Manuscr. Math. **47**, 133–151 (1984)

[A4]   G. Aronsson, Construction of singular solutions to the p-harmonic equation and its limit equation for $p = \infty$. Manuscr. Math. **56**, 135–158 (1986)

[A5]   G. Aronsson, On certain p-harmonic functions in the plane. Manuscr. Math. **61**, 79–101 (1988)

[A6]   G. Aronsson, Representation of a p-harmonic function near a critical point in the plane. Manuscr. Math. **66**, 73–95 (1989)

[Al]   G. Alessandrini, Critical points of solutions to the p-Laplace equation in dimension two. Boll. Della Unione Mat. Ital. Sez. A Ser. VII **1**, 239–246 (1987)

[AL]   G. Aronsson, P. Lindqvist, On p-harmonic functions in the plane and their stream functions. J. Differ. Equ. **74**, 157–178 (1988)

[ALl]   Á. Arroyo, J. Llorente, On the asymptotic mean value property for planar p-harmonic functions. Proc. Am. Math. Soc. **144**, 3859–3868 (2016)

[AS]   S. Armstrong, Ch. Smart, An easy proof of Jensen's theorem on the uniqueness of infinity harmonic functions. Calc. Var. Part. Differ. Equ. **37**, 381–384 (2010)

[B]   M. Brelot, *Éléments de la Théorie Classique du Potentiel*, 2e edn. (Paris, 1961)

[Be]   L. Bers, *Mathematical Aspects of Subsonic and Transonic Gas Dynamics, Surveys in Applied Mathematics III* (Wiley, New York, 1958)

[BB]   G. Barles, J. Busca, Existence and comparison results for fully nonlinear degenerate elliptic equations without zeroth-order term. Commun. Part. Differ. Equ. **26**, 2323–2337 (2001)

[BDM]   T. Bhattacharya, E. DiBenedetto, J. Manfredi, Limits as $p \to \infty$ of $\Delta_p u = f$ and related problems. Rendiconti del Seminario Matematico Università e Polytecnico di Torino 15–68 (1989)

[BG]   E. Bombieri, E. Giusti, Harnack's inequality for elliptic differential equations on minimal surfaces. Inven. Math. **15**, 24–46 (1972)

[BI1]   B. Bojarski, T. Iwaniec, *p-Harmonic Equation and Quasiregular Mappings. Partial Differential Equations* (Warsaw, 1984), pp. 25–38. Banach Center Publications (19) (1987)

© The Author(s), under exclusive license to Springer Nature Switzerland AG 2019   101
P. Lindqvist, *Notes on the Stationary p-Laplace Equation*,
SpringerBriefs in Mathematics, https://doi.org/10.1007/978-3-030-14501-9

[BI2]   B. Bojarski, T. Iwaniec, Analytical foundations of the theory of quasiconformal mappings in $\mathbf{R}^n$. Ann. Acad. Sci. Fenn. Ser. AI **8**, 257–324 (1983)

[Br]   K. Brustad, Superposition in the p-Laplace equation. Nonlinear Analysis. Theory Methods Appl. **158**, 23–31 (2017)

[CEG]   M. Crandall, L. Evans, R. Gariepy, Optimal Lipschitz extensions and the infinity Laplacian. Calc. Var. Part. Differ. Equ. **13**, 123–129 (2001)

[CL]   H. Choe, J. Lewis, On the obstacle problem for quasilinear elliptic equations of p-Laplacian type. SIAM J. Math. Anal. **22**, 623–638 (1991)

[CIL]   M. Crandall, H. Ishii, P.-L. Lions, User's guide to viscosity solutions of second order partial differential equations. Bull. Am. Math. Soc. **27**, 1–67 (1992)

[CZ]   M. Crandall, J. Zhang, Another way to say harmonic. Trans. Am. Math. Soc. **355**, 241–263 (2003)

[D]   B. Dacorogna, *Direct Methods in the Calculus of Variations* (Springer, Heidelberg, 1989)

[Db1]   E. Di Benedetto, $C^{1,\alpha}$ local regularity of weak solutions of degenerate elliptic equations. Nonlinear Anal. **7**, 827–850 (1983)

[Db2]   E. DiBenedetto, *Degenerate Parabolic Equations* (Springer, New York, 1993)

[Dg]   E. De Giorgi, Sulla differenziabilità e l'analiticità delle estremali degli integrali multipli regolari. Mem. Accad. Sci. Torino (Classe di Sci.mat., fis. e nat.) **3**(3), 25–43 (1957)

[E]   L. Evans, A new proof of local $C^{1,\alpha}$ regularity for solutions of certain degenerate elliptic P.D.E. J. Differ. Equ. **45**, 356–373 (1982)

[EG]   L. Evans, R. Gariepy, *Measure Theory and Fine Properties of Functions* (CRC Press, Boca Raton, 1992)

[ES]   L. Evans, Ch. Smart, Everywhere differentiability of infinity harmonic functions. Calc. Var. Part. Differ. Equ. **42**, 289–299 (2011)

[F]   M. Fuchs, p-harmonic obstacle problems. III. Boundary regularity, Annali di Matematica Pura ed Applicata (Series IV) **156**, 159–180 (1990)

[GZ]   R. Gariepy, W. Ziemer, *Modern Real Analysis* (PWS Publishing Company, Boston, 1994)

[G]   E. Giusti, *Metodi Diretti nel Calcolo delle Variazioni* (UMI, Bologna 1994). English translation: *Direct Methods in the Calculus of Variations* (World Scientfic Publ. Co., River Edge, 2003)

[GLM]   S. Granlund, P. Lindqvist, O. Martio, Note on the PWB-method in the nonlinear case. Pac. J. Math. **125**, 381–395 (1986)

[GT]   D. Gilbarg, N. Trudinger, *Elliptic Partial Differential Equations of Second Order*, 2nd edn. (Springer, Heidelberg, 1983)

[HF]   G. Hong, X. Feng, Superposition principle on viscosity solutions of infinity Laplace equation. Nonlinear Anal. **171**, 32–40 (2018)

[HL]   Q. Han, F. Ling, *Elliptic Partial Differential Equations (Courant Lecture Notes)* (New York, 1997)

[I]   T. Iwaniec, G. Martin, Quasiregular mappings in even dimensions. Acta Math. **170**, 29–81 (1993)

[IM]   T. Iwaniec, J. Manfredi, Regularity of p-harmonic functions in the plane. Rev. Mat. Iberoam. **5**, 1–19 (1989)

[J]   R. Jensen, Uniqueness of Lipschitz extensions minimizing the sup-norm of the gradient. Arch. Rat. Mech. Anal. **123**, 51–74 (1993)

[JLM]   P. Juutinen, P. Lindqvist, J. Manfredi, On the equivalence of viscosity solutions and weak solutions for a quasilinear equation. SIAM J. Math. Anal. **33**, 699–717 (2001)

[JN]   F. John, L. Nirenberg, On functions of bounded mean oscillation. Commun. Pure Appl. Matmematics **14**, 415–426 (1961)

[Jo]   J. Jost, *Partielle Differentialgleichungen* (Springer, Heidelberg, 1998)

[JJ]   V. Julin, P. Juutinen, A new proof of the equivalence of weak and viscosity solutions for the p-Laplace equation. Commun. Part. Differ. Equ. **37**, 934–946 (2012)

[KMP]   B. Kawohl, J. Manfredi, M. Parviainen, Solutions of nonlinear PDEs in the sense of averages. Journal de Mathématiques Pures et Appliquées **97**(9), 173–188 (2012)

[KV]   S. Kichenassamy, L. Véron, Singular solutions of the p-Laplace equation. Math. Ann.
          275, 599–615 (1986)

[KKT]   T. Kilpeläinen, T. Kuusi, A. Tuhola-Kujanpää, Superharmonic functions are locally renor-
          malized solutions. Annales de l'institut Henri Poincaré, Analyse Non Linéaire **28**, 775–
          795 (2011)

[KM1]   T. Kilpeläinen, J. Malý, Degenerate elliptic equations with measure data and nonlinear
          potentials, Annali della Scuola Normale Superiore di Pisa (Science Fisiche e Matem-
          atiche). Serie IV **19**, 591–613 (1992)

[KM2]   T. Kilpeläinen, J. Malý, The Wiener test and potential estimates for quasilinear elliptic
          equations. Acta Math. **172**, 137–161 (1994)

[K]   S. Koike, *A Beginner's Guide to the Theory of Viscosity Solutions (MSJ Memoirs 13*
          (Mathematical Society of Japan, Tokyo, 2004)

[KM]   T. Kuusi, G. Mingione, Linear potentials in nonlinear potential theory. Arch. Rat. Mech.
          Anal. **207**, 215–246 (2013)

[Le1]   J. Lewis, Capacitary functions in convex rings. Arch. Rat. Mech. Anal. **66**, 201–224
          (1977)

[Le2]   J. Lewis, Regularity of the derivatives of solutions to certain degenerate elliptic equations.
          Indiana Univ. Math. J. **32**, 849–858 (1983)

[Le3]   J. Lewis, On very weak solutions of certain elliptic systems. Commun. Part. Differ. Equ.
          18, 1515–1537 (1993)

[L1]   P. Lindqvist, On the growth of the solutions of the equation div $(|\nabla u|^{p-2}\nabla u) = 0$ in
          n-dimensional space. J. Differ. Equ. **58**, 307–317 (1985)

[L2]   P. Lindqvist, On the definition and properties of p-superharmonic functions. Journal für
          die reine und angewandte Mathematic (Crelles Journal) **365**, 67–79 (1986)

[L3]   P. Lindqvist, Regularity for the gradient of the solution to a nonlinear obstacle problem
          with degenerate ellipticity. Nonlinear Anal. **12**, 1245–1255 (1988)

[L4]   P. Lindqvist, *Notes on the Infinity Laplace Equation (SpringerBriefs in Mathematics)*
          (Springer, 2016)

[LiM1]   P. Lindqvist, J. Manfredi, Viscosity solutions of the evolutionary p-Laplace equation.
          Differ. Integr. Equ. **20**, 1303–1319 (2007)

[LiM2]   P. Lindqvist, J. Manfredi, On the mean value property of the p-Laplace equation in the
          plane. Proc. Am. Math. Soc. **144**, 143–149 (2016)

[LiM3]   P. Lindqvist, J. Manfredi, Note on a remarkable superposition for a nonlinear equation.
          Proc. Am. Math. Soc. **136**, 133–140 (2018)

[LM]   P. Lindqvist, O. Martio, Two theorems of N. Wiener for solutions of quasilinear elliptic
          equations. Acta Math. **155**, 153–171 (1985)

[LU]   O. Ladyzhenskaya, N. Uraltseva, *Linear and Quasilinear Elliptic Equations* (Academic
          Press, New York, 1968)

[M1]   J. Manfredi, p-harmonic functions in the plane. Proc. Am. Math. Soc. **103**, 473–479
          (1988)

[M2]   J. Manfredi, Isolated singularities of p-harmonic functions in the plane. SIAM J. Math.
          Anal. **22**, 424–439 (1991)

[MPR1]   J. Manfredi, M. Parviainen, J. Rossi, An asymptotic mean value characterization for
          p-harmonic functions. Proc. Am. Math. Soc. **138**, 881–889 (2010)

[MPR2]   J. Manfredi, M. Parviainen, J. Rossi, On the definition and properties of p-harmonious
          functions. Annali della Scuola Normale Superiore di Pisa (Classe di Scienze), Serie V
          11, 215–241 (2012)

[MPR3]   J. Manfredi, M. Parviainen, J. Rossi, Dynamic programming principle fot tug-of-wae
          games with noise. ESAIM Control. Optim. Calc. Var. **18**, 81–90 (2012)

[MW]   J. Manfredi, A. Weitsman, On the Fatou theorem for p-harmonic functions. Commun.
          Part. Differ. Equ. **13**, 651–668 (1988)

[M]   O. Martio, Counterexamples for unique continuation. Manuscr. Math. **60**, 21–47 (1988)

[Ma]   V. Maz'ja, On the continuity at a boundary point of solutions of quasilinear elliptic
          equations. Vestnik Leningradskogo Universiteta **13**, 42–55 (1970). (in Russian)

[Mo1]   J. Moser, A new proof of De Giorgi's theorem concerning the regularity problem for elliptic differential equations. Commun. Pure Appl. Math. **13**, 457–468 (1960)

[Mo2]   J. Moser, On Harnack's theorem for elliptic differential equations. Commun. Pure Appl. Math. 577–591 (1961)

[MZ]   J. Michel, P. Ziemer, Interior regularity for solutions to obstacle problems. Nonlinear Anal. **10**, 1427–1448 (1986)

[PSW]   Y. Peres, O. Schramm, S. Sheffield, D. Wilson, Tug-of-war and the infinity Laplacian. J. Am. Math. Soc. **22**, 167–210 (2009)

[R]   T. Radó, *Subharmonic Functions* (New York, 1949)

[Re]   J. Reshetnyak, Extremal properties of mappings with bounded distortion. Sibirskij Matematicheskij Zhurnal **10**, 1300–1310 (1969). (in Russian)

[Ro]   J. Rossi, Tug-of-war games and PDEs. Proc. R. Soc. Edinb. Sect. Math. **141**, 319–369 (2011)

[S]   S. Sakaguchi, Coincidence sets in the obstacle problem for the p-harmonic operator. Proc. Am. Math. Soc. **95**, 382–386 (1985)

[Sa]   O. Savin, C^1 regularity for infinity harmonic functions in two dimensions. Arch. Rat. Mech. Anal. **176**, 351–361 (2005)

[SC]   L. Saloff-Coste, *Aspects of Sobolev-Type Inequalities*, London Mathematical Society Lecture Note Series 289 (Cambridge, 2002)

[T1]   N. Trudinger, On Harnack type inequalities and their application to quasilinear elliptic equations. Commun. Pure Appl. Math. **20**, 721–747 (1967)

[T2]   N. Trudinger, On the regularity of generalized solutions of linear, non-uniformly elliptic equations. Arch. Rat. Mech. Anal. **42**, 50–62 (1971)

[To]   P. Tolksdorf, Regularity for a more general class of quasilinear elliptic equations. J. Differ. Equ. **51**, 126–150 (1984)

[Uh]   K. Uhlenbeck, Regularity for a class of nonlinear elliptic systems. Acta Math. **138**, 219–240 (1977)

[Ur]   N. Ural'ceva, Degenerate quasilinear elliptic systems, Zap. Naučn. Sem. Leningrad. Otdel. Mat. Inst. Steklov **7**, 184–192 (1968). (in Russian)

[Wi]   K.-O. Widman, Hölder continuity of solutions of elliptic equations. Manuscr. Math. **5**, 299–308 (1971)

[Y]   Y. Yu, A remark on C^2 infinity-harmonic functions. Electron. J. Differ. Equ. **2006**(122), 1–4 (2006)

Printed in the United States
By Bookmasters